Social Behaviour in Animals

Originally published in 1953, this is a classic study in animal behaviour, drawing on the author's own extraordinary studies of insects, fish, and birds, as well as on the literature. The concept 'community' is taken in its widest sense to include all types of association of individuals, not only flocks and herds, but also the family, the pair, and even two animals engaged in combat.

The author received the Nobel Prize for his work in this field in 1973.

Social Behaviour in Animals

Social Behaviour
in Animals

With special reference to vertebrates

N. Tinbergen

LONDON AND NEW YORK

First published 1953 by Methuen & Co. Ltd
Second edition 1964

This edition first published in 2014 by Psychology Press

Published 2022 by Routledge
2 Park Square, Milton Park, Abingdon, Oxon OX14 4RN
605 Third Avenue, New York, NY 10017

Routledge is an imprint of the Taylor & Francis Group, an informa business

A Library of Congress record exists under LC control no.: 67006130

ISBN: 978-1-848-72297-2 (hbk)
ISBN: 978-1-315-84999-7 (ebk)

Social Behaviour in Animals

WITH SPECIAL REFERENCE TO VERTEBRATES

N. TINBERGEN

Reader in Animal Behaviour in the University of Oxford

LONDON

CHAPMAN AND HALL

First published 1953
by Methuen & Co. Ltd
Reprinted three times
Second edition 1964
Reprinted 1965, 1969

First published as a Science Paperback 1965
by Chapman and Hall Ltd
11 *New Fetter Lane, London EC4P 4EE*
Reprinted 1966, 1968, 1969, 1971, 1972, 1975

© 1964 *N. Tinbergen*

Printed in Great Britain by
Butler & Tanner Ltd
Frome and London

ISBN 0 412 20000 7

PREFACE

THIS book is not intended as an exhaustive review of facts. Its aim is rather the presentation of a biological approach to the phenomena of social behaviour. This type of approach was revived by Lorenz's pioneer studies. It is characterized by emphasis on the need for renewed and careful observation of the huge variety of social phenomena occurring in nature; by emphasis on a balanced study of the three main biological problems—function, causation, evolution; by emphasis on an appropriate sequence of description, qualitative analysis, quantitative analysis; and finally by emphasis on the need for continuous re-synthesis.

The character of this approach, combined with the limitations of space, have determined this book's contents. Limits of space led to the omission of a great deal of description. Thus, Deegener's voluminous work on the multitude of types of animal aggregations has not been discussed. Also, the highly specialized 'states' of social insects have not been treated in detail, since there are excellent books dealing exclusively with them.

The nature of the approach makes this book essentially different from other books on social behaviour. On the one hand, I have treated briefly some problems which have been much more elaborated by other authors. Thus Allee's works are mainly concerned with the various uses animals derive from crowding; there is little mention of the causes underlying social co-operation, and when dealing with these causes, attention is focused entirely on the phenomenon of peck-order—an interesting, but minor aspect of social organization. Other workers seem to attach undue value to the influence of transmission of food from one individual to another; while this is admittedly a factor in the development of some social relationships, it is again merely one element of a large complex of phenomena. Lastly, there is an enormous amount of scattered and often unrelated analytical evidence, acquired under such special laboratory conditions that it is at present impossible to say how it is related to the normal life of the species concerned.

On the other hand, I considered it of great importance to work out the formulation of the main problems, of their relation to each other and to more special, subordinated problems. This task, together with the necessary descriptions of many new facts found through 'naturalistic' study, and with the first qualitative steps of analysis, required much space. In addition, I wanted to formulate and emphasize some new theories which I consider important because of their great heuristic value. Thus the significance of intraspecific fighting, the causation of threat and courtship behaviour, the functions of releasers, and other problems to which the new approach has made distinct contributions, have been presented in some detail, and an attempt has been made to give them their proper place in the complex system of problems.

I have tried to present my thoughts in such a way that they can easily be followed by interested non-professionals. It is my hope that by doing so I will stimulate research, for I am convinced that the amateur can still make great contributions to our young science.

I am much indebted to Dr. Michael Abercrombie and to Desmond Morris for valuable criticism and for revising the English text; to Dr. L. Tinbergen for drawing part of the illustrations, and to the Oxford University Press for permission to use a number of the illustrations from my book *The Study of Instinct*. My thanks are further due to Dr. Hugh Cott for permission to reproduce Fig. 61 and to Dr. Brian Roberts for permission to use his splendid penguin photograph used for the wrapper and on Plate 5.

CONTENTS

TEXT ILLUSTRATIONS

PLATES

CHAPTER I

INTRODUCTION

STATEMENT OF THE PROBLEMS

SINCE Starlings living in flocks are called social, whereas a Peregrine Falcon, hunting above the estuaries in winter, is clearly solitary, 'social' indicates that we have to do with more than one individual. There need not be many individuals; I would even call much in the behaviour of a pair of animals 'social'.

Not all aggregations of animals however are social. When, on a summer night, hundreds of insects gather round our lamp, these insects need not be social. They may have arrived one by one, and their gathering just here may be clearly accidental; they aggregate because each of them is attracted by the lamp. But Starlings on winter evenings, executing their fascinating aerial manœuvres before settling down for the night do really react to one another; they even follow each other in such perfect order that we may be led to believe that they have superhuman powers of communication. This keeping together on the basis of reacting to each other, then, is another mark of social behaviour. In this respect, animal sociology differs from plant sociology, which includes all phenomena of plants occurring together, irrespective of whether they influence each other or are merely attracted in the same way by the same external agents.

The influence which social animals exert on each other is not merely attraction. Aggregation is usually but the mere prelude to closer co-operation, to doing something together. In the case of the Starlings this co-operation is simple; they just fly around together, they execute the same turnings, some may utter alarm-calls to which others react; they may join in warding off a Sparrow Hawk or a Peregrine Falcon by clustering together and rising above the predator. The coming-together of a male and a female in the breeding season may be followed by a long period of close and intricate co-operation, in mating, in nest-building, in incubation, in rearing the young.

The study of social behaviour therefore is the study of co-operation between individuals. There may be two individuals involved, or more. In the Starling flock thousands of individuals may co-operate.

When we speak of co-operation, we have always at the back of our minds an idea, clear or vague, of the purpose of this co-operation. We assume that it serves some end. This problem of the 'biological significance', or 'function' of life processes is one of the attractive problems of biology. It exists in the physiology of the individual, and also in that of one of its organs. On the other hand, proceeding to a higher level of integration, it exists in sociology. Whereas the physicist or the chemist is not intent on studying the purpose of the phenomena he studies, the biologist has to consider it. 'Purpose' of course is meant here in a restricted sense. I do not mean that the biologist is more concerned with the problem of why there should be life at all, than the physicist with the problem of why there should be matter and movement at all. But the very nature of living things, their unstable state, leads us to ask: how is it possible that living things do not succumb to the omnipresent destructive influences of the environment? How do living things manage to survive, to maintain and to reproduce themselves? The purpose, end, or goal of life processes in this restricted sense is maintenance, of the individual, of the group, and of the species. A community of individuals has to be kept going, has to be protected against disintegration just as much as an organism, which, as its name implies, is a community of parts—of organs, of parts of organs, of parts of parts of organs. Just as the physiologist asks how the individual, or the organ, or the cell, manages to maintain itself by organized co-operation of its constituents, so the sociologist has to ask how the constituents of the group—the individuals—manage to maintain the group.

In this chapter I will first, by way of reconnaissance, give a number of examples of group life in various animal species. I will then, in subsequent chapters, proceed to examine which functions are served by the social behaviour of the constituent individuals to the benefit of other individuals or of the group as a whole. Next I will consider how co-operation is organized. These two aspects, that of the function and that of the causation of social behaviour, will be discussed for the several types of social behaviour: the behaviour of sex partners, family and group life, and fighting. In this way we will discover, step by step, social structures. Since these social structures are almost always temporary structures, we will have to study how they arise. Finally we must try to find out how organisms in the course of their long evolution have developed the type of social organization we observe to-day.

THE HERRING GULL (*Larus argentatus*) [25], [71], [105]

All through autumn and winter Herring Gulls live in flocks. They feed in flocks, migrate in flocks, sleep in flocks. When you watch foraging Herring Gulls from day to day, you will find that it is usually not a common reaction to an outside agent such as abundant food that brings them together. One group of Herring Gulls I have known used to catch earthworms in the meadows. I would find them on one meadow one day, on another meadow next day. Now and then the whole flock moved from one place to another. Earthworms were plentiful in all these places, and there was not the slightest indication that the gulls moved because the earthworm supply was locally exhausted. Decimating an earthworm population is not that easy! Whenever individual gulls came from other feeding grounds, they invariably made for the flock and did not settle at any other place in the meadow. It was the other gulls that attracted them.

The gulls in the flock react to each other in various ways. When you approach them too closely, some of them stop feeding, crane their necks, and look intently at you. Soon others do the same, and the whole flock stands staring at you. One may then utter the alarm-call, a rhythmic 'ga-ga-ga', and all at once it flies off. Immediately the others follow, and the whole company leaves. The reaction is almost simultaneous. This might of course be due to their simultaneously reacting to us, the

outside agents releasing this behaviour. But often enough, for instance when you stalk them under cover, only one or two birds
may discover you, and then you can see how their behaviour
—stretching the neck, or calling, or flying off suddenly—influences the others, who may not have perceived the danger
themselves.

In spring the flock visits the breeding grounds in the sand
dunes together. When they settle, after having circled in the
air above for some time, they segregate into pairs, which settle
on territories within the colony's range. Not all birds are paired
however; many gather in so-called 'clubs'. Long and consistent
study of marked individuals has proved that new pairs form in
these clubs. The females take the initiative in pair formation.

FIG. 1.—Male Herring Gull (*left*) about to feed female

An unmated female approaches a male in a peculiar attitude.
She withdraws her neck, points her bill forward and slightly
upward, and, adopting a horizontal attitude, she walks slowly
round the male of her choice. The male may react in one of two
possible ways. Either he begins to strut around and attacks
other males, or he may utter a long-drawn call and walk away
with the female. The female then often begins to beg for food
by making curious tossing movements with her head. The male
responds to this begging behaviour by regurgitating some food,
which the female greedily devours (Fig. 1). In the beginning of
the season this may be a mere flirtation, and no serious bond
need arise from it. But usually such pairs begin to be attached
to each other, and in this way pair formation takes place. Once
the pair is formed, the next step is taken: they go house-hunting.
They leave the club, and select a territory somewhere in the

colony. Here they begin to build a nest. Both partners collect nest material and carry it to the selected nest-site. There they sit down in turn, mould a sort of shallow pit by scraping movements of the legs, and line it with grass and moss.

Once or twice a day the birds copulate. This is always introduced by a lengthy ceremony. Either one of the mates begins to toss its head, as if it is begging for food. The difference with 'courtship feeding' is that both birds make these head-tossing movements. They go on for quite a while, and gradually the male begins to stretch its neck, and soon it jumps up into the air and mounts the female. Copulation is then effected by the male bringing its cloaca repeatedly in touch with that of the female.

Coinciding with pair formation, nest building, courtship feeding, and copulation, another behaviour pattern has appeared, particularly in the male: fighting. Already in the club the male's aggressiveness can be so intense that it chases away all the gulls in the vicinity. Once established on the territory the male becomes entirely intolerant of trespassers. Each intruding male is attacked. Usually no genuine attack is made, threat alone is often sufficient to drive a stranger off. There are three types of threat. The mildest form is the 'upright threat posture': the male stretches its neck, points its bill down, and sometimes lifts its wings (Fig. 2). In this attitude it walks towards the stranger in a remarkably stiff way, all its muscles tense. A stronger expression of the same intention is 'grass pulling'.

FIG. 2. — Upright threat posture of male Herring Gull

The male walks up till quite close to the opponent, and all at once bends down and pecks furiously into the ground. It takes hold of some grass, or moss, or roots, and pulls it out. When male and female face a neighbouring pair together, they show a third type of threat: 'choking'. They bend their heels, lower the breast, point their beaks down and with lowered tongue bones, which give them a very curious facial expression,

2

they make a series of incomplete pecking movements at the ground. This is accompanied by a rhythmic, hoarse sort of cooing call.

All these threat movements obviously impress other gulls. They understand the aggressive meaning, and often retreat.

When the eggs are laid, the pair take turns in sitting on them.

Here again their co-operation is very impressive. They never leave the eggs alone. While one is incubating, the other may be feeding miles away. When it comes back, the sitting bird waits until the newcomer walks up to the nest. This approach is accompanied by special movements and calls. Usually the long-drawn 'mew call' is uttered, and often some nest material is carried. Then the sitting bird rises, and the other takes its place.

The care of the eggs might be called social behaviour, for from the time of being laid the eggs are individuals. Usually we do not consider such one-sided relations as really social, but we must not forget that the egg, although not moving, does give special stimuli which have a profound influence on the parent bird.

As soon as the eggs hatch, however, the relationships between parents and offspring become truly mutual. In the beginning the chicks are not doing much except passively being brooded, but after a few hours they begin to beg for food. When the

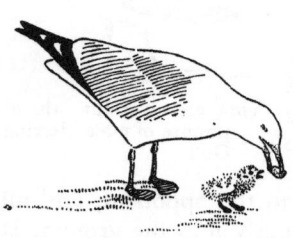

FIG. 3.—Herring Gull feeding chick

parent gives them the opportunity by getting up, the chicks begin to aim a series of pecking movements directed at the parent's bill tip. Soon the latter regurgitates food: a half-digested fish, or a crab, or a batch of earthworms. It takes a small morsel between the bill tips and patiently offers it to the chicks (Fig. 3), waiting with forward-bent head until one of them, after various failures, manages to get hold of it and swallows it. A new titbit is presented, and perhaps several more. The chicks then stop begging, the parent quickly swallows the remains again, and settles down once more to brood.

Another relationship between parents and offspring becomes apparent when predators enter the colony. Dogs, Foxes and Humans evoke the most intense reactions. The adults utter the well-known alarm-call 'gagaga! gagagagaga!' and fly up. This call acts as a communication in two ways. The chicks run for cover, and crouch. The adult members of the colony all fly up, and prepare for attack. Actual attacks on the intruder however are done by the pairs individually. Each bird will swoop down, and may even hit the predator with one or both legs, when the latter comes near the nest. Occasionally such an attack is accompanied by 'bombing' with regurgitated food or faeces; a very distasteful weapon. Yet such attacks do not have complete success. Foxes and Dogs, and also Humans, are somewhat disturbed and distracted by them, and are certainly prevented from searching as thoroughly as they would if left alone. They may miss some nests, and especially young, in this way, but it does not prevent them from finding those upon which they stumble by accident. This relative inefficiency however is found in all biological functions: none of them leads to absolute and complete success. However, they all contribute something towards success. Of great help in the defence against predators is the cryptic colour and behaviour of the young. As a matter of fact, the whole function of crouching (Fig. 4) is to avoid catching the eye of a visually hunting predator.

FIG. 4.—Herring Gull chick crouching

After a day or so, the chicks become more mobile. They crawl around on the territory, gradually moving further away from the nest. They do not leave the territory, however, unless compelled to do so by frequent human disturbance, such as visits of crowds of nature lovers. Too much love is often fatal to the chicks, for when they leave the territory they are attacked and often killed by the neighbours. The true nature lover might get more satisfaction from patiently watching the gulls' life from a distance. Most of the happenings described here can then be watched.

Thus we see numerous proofs of social organization. Part of this organization serves the purpose of mating. There is co-operation between male and female, however, which has nothing to do with mating, but serves the family. Beyond that there is co-operation between parents and chicks. The chicks urge the parents to feed them; the parents may urge the chicks to hide and keep still. There is also co-operation between different pairs; the alarm-call raises the whole colony. The result is the rearing of large numbers of young birds, a result to which we are so much used that it seems commonplace to mention it; yet even minor disturbances of the intricate social pattern may be fatal. To mention just one such disturbance: I observed several times that an incubating gull got up to 'stretch its legs' for a minute. As it stood preening about two yards from the nest, another gull swooped down and pecked at one of the eggs, breaking it right in two halves. Before it could begin to eat its contents, the parent gull chased it away. Yet the egg was lost through the parent's negligence. Another case: in one pair which I watched, the male had no brooding urge. It never relieved the female. She persevered heroically and sat on the eggs almost without break for twenty days. On the twenty-first day she deserted, and the brood was lost. However disastrous this was for the young, it was a blessing for the species, for what if the offspring inherited this defect from the father and supplied the species with three instead of one of these degenerates?

THE THREE-SPINED STICKLEBACK
(*Gasterosteus aculeatus*) [50, 51, 70, 101, 110]

Outside the breeding season, Sticklebacks live in schools. When they are foraging together, we may watch one type of behaviour which we did not notice so much in the gulls, although they do show it too. When one fish happens to find a particularly satisfying titbit, and starts to devour it in the greedy Stickleback manner, others rush towards it and try to rob it. This may have a partial result, for some may manage to tear the prey to pieces and thus secure a share. Others are less lucky, and then they start to search at the bottom. This means that when and wherever one member of the school finds food, others may be stimulated to search then and there, and in this way congregations of prey animals are liable to be discovered and utilized to the last.

As in the Herring Gull, the breeding season starts a much more complicated system of social co-operation than we ever see in autumn or winter. To begin with, the males isolate themselves from the school, and select territories. They assume brilliant nuptial colours. The eye becomes a shining blue, the back, instead of dull brownish, becomes greenish, the underparts become red. Whenever another fish, and particularly another male, enters the territory, it is attacked (Fig. 5).

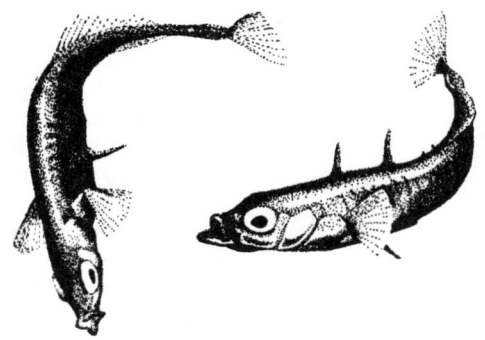

FIG. 5.—Boundary fight of two Three-spined Stickleback males
(*after Ter Pelkwijk and Tinbergen, 1937*)

Again, fighting is rarer than threat. The threat behaviour of male Sticklebacks is peculiar. Not only do they dart towards the opponent with raised dorsal spines and opened mouth, ready to bite, but, when the opponent does not flee at once but resists, the owner of the territory does not actually bite but points its head down and, standing vertically in the water, makes some jerky movements as if it were going to bore its snout into the sand. Often it erects one or both ventral spines.

When the male is undisturbed it begins to build a nest. It selects a site, and here it begins to take up mouthfuls of sand from the bottom, carries them away and drops the loads some five or six inches away. In this way a shallow pit is made. Then

FIG. 6.—Courtship sequence of Three-spined Stickleback (*after Tinbergen, 1951*)

the male gathers nest material, usually threads of algae, and presses them down into the pit. Occasionally it creeps with slow quivering movements over the material, secreting a sticky glue from the kidneys which pastes the plants together. In the course of some hours or days a kind of green cluster results, through which the male then bores a tunnel by wriggling itself right through.

The nest is now finished. At once the male changes its colour. The red becomes still more intense, and all the black colour cells which are found on the back contract to minute dots. Thereby the underlying glittering bluish crystals of guanin which are situated in a deeper stratum of the skin are exposed, and the back now becomes a shiny whitish blue. The light back and the dark red underside, together with the brilliant eye, now make the male extremely conspicuous. Displaying this attractive dress, the male parades up and down its territory.

In the meantime the females, which have not bothered about

nest building at all, have developed a brilliant silvery gloss, and their bodies are heavily swollen by the bulky eggs which have developed in the ovaries. They cruise about in schools. In a good Stickleback habitat, they pass through occupied territories repeatedly during the day. Each male, if ready to receive a female, reacts to them by performing a curious dance towards and all around them (Fig. 6). Each dance consists of a series of leaps, during which the male first turns as if going to swim away from the females, then abruptly turns towards them with its mouth wide open. Sometimes it may hit a female, but usually it stops just in front of it, and then turns away for a new performance. This zigzag dance frightens most of the females away, but a single one may be sufficiently matured to be willing to spawn, and such a female does exactly the opposite from fleeing: it turns towards the male, at the same time adopting a more or less upright attitude. The male now immediately turns round and swims hurriedly towards the nest. The female follows it. Arrived at the nest, the male thrusts its snout into the entrance, turning along its body axis, so that it lies on its side, its back towards the female, which now tries to wriggle into the nest. With a strong beat of her tail she manages to penetrate into the narrow opening, and slips in. She remains in the nest, her head protruding from one end, the tail from the other. The male now begins to prod her tail base with his snout, giving series of quick thrusts. After some time the female begins to lift her tail, and soon she spawns. This done, she quietly pushes through the nest, while the male enters it and, slipping through in his turn, fertilizes the eggs. Then he chases her away, returns to the nest, restores the roof which has been lifted and torn by the passage of the two, and often he adjusts and shifts the eggs, concealing them well under the roof. This is the end of the whole mating ceremony. There is no 'marriage', no personal relationships, and the female's whole task in reproduction is just to provide the eggs. The whole care of eggs and young is the male's job. The association between male and female therefore is just a series of quick reciprocal reactions, which can be summed up as shown on the next page.

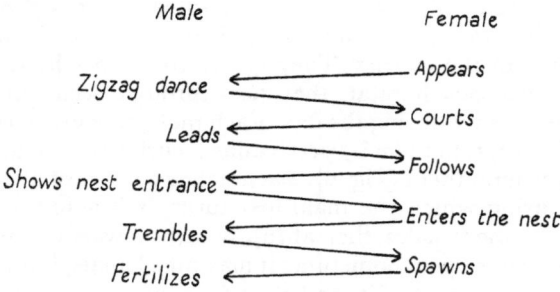

The male may court two, three, or even more females in the course of a few days and thus collect several clutches of eggs in its nest. Then its sex drive wanes, and instead of courting it begins to show parental behaviour. This consists of warding off intruders, whether males, females, other fish, or predators; and of ventilation of the eggs. This is done by a remarkable movement, called 'fanning'. Standing in front of the nest entrance, its head pointing obliquely down, the male sends a water current down on to the nest by alternate forward movements of the pectoral fins. To counteract the backward pressure this exerts upon his body, he makes forward swimming movements with his tail, which keep him exactly at the same spot. Through these movements water is sucked towards him mainly from above and from below, and sent partly towards the nest, partly to the rear. Complicated stimulus situations from the nest, the environment, and the eggs control this activity. The time spent in fanning increases in the course of the next eight days. In the beginning, about 200 seconds may be given to fanning each half-hour. This gradually increases until at the end of the week fanning may take up three-quarters of his time. This increase is partly due to a growing internal drive, partly to increasing stimulation by the eggs, which consume increasing amounts of oxygen as development proceeds: it is the resulting lack of oxygen which activates the male's fanning.

The young hatch from the eggs after seven or eight days, but remain in the nest for another day or so. Then they begin to move about. When this happens, the male's fanning stops

rather abruptly, and it now guards the young carefully (Fig. 7).
As soon as one begins to swim or rather wriggle away from the
school, the male snaps it up in its mouth, and spits it back into
the swarm. The young are usually too sluggish to be able to
escape. On one occasion however they manage to get away
from the male: one by one you can watch them suddenly shoot
up towards the surface of the water, touch it, and then dash
down again. The male often sees it, tries to follow them, but
always misses them, and only succeeds in capturing them after

FIG. 7.—Male Three-spined Stickleback guarding young

they have come down again. This curious behaviour of the
young has a special function: at the surface they snap up a tiny
air bubble, which is passed on through the gut and a narrow
side tube to the swim bladder. Once this initial air bubble has
arrived there, the bladder can proceed to make more gas by
itself. The dashing excursion to the surface, which each young
fish performs once in its life, has to be so quick for two reasons:
escape from possible predators, and escape from the good
intentions of the male.

In the course of the next two weeks, the young develop more
and more initiative and move further and further from the

nest. The male's tendency to keep them together wanes, but instead the young keep together of their own accord. Yet the male still guards them. Gradually however he loses interest, as well as his brilliant colour, and after some weeks he leaves the territory and seeks the company of his colleagues, while the young keep to companions of their own age.

Thus the social behaviour of the Sticklebacks resembles that of the Herring Gull in many respects. There is co-operation of male and female towards the end of fertilizing the eggs, although in the Stickleback the association of the two sexes does not go beyond that. There are relations between the male and the eggs, between the male and its young, between the young themselves, and there are fights. The young stimulate the father in certain ways, and he responds by various types of parental conduct. Whether the father influences the young (apart from carrying them back now and then) and thus forces them to stay near the nest, is still uncertain.

THE GRAYLING (*Eumenis semele*) [108]

Let us now study the behaviour of an insect. I take the Grayling Butterfly (Fig. 8) because I happen to know its behaviour better than that of other insects.

The caterpillars spend the autumn and the winter on the hard dry grasses which grow in the arid habitat of this species. They pupate at the end of the spring. In the beginning of July the butterflies begin to emerge. They spend part of their time feeding. They suck nectar on various flowers and they visit 'bleeding' trees, particularly such that have been damaged by the larvae of the Goat Moth, which bore in live wood. The Graylings may be seen in groups of five, ten, or even more, but these gatherings are not at all social; this is just another case of attraction by outside stimuli: the colour and scent of the food. Soon the reproductive behaviour patterns appear. The males stop foraging and take up stations on the ground, or on the bark of trees. They are very much on the alert, and whenever another butterfly passes, they fly up and pursue it. If the passing butterfly is a female Grayling ready to mate, she reacts to the male's approach by alighting on the ground. The male follows her and alights behind her. He then walks up towards her and takes up a position in front of, and facing her. If she does not respond by wing-flapping (which indicates a suboptimal mating drive and which often drives the male away), but keeps motionless, he will begin his elegant courtship. First, he jerks his wings upward and forward a few times in quick succession. Then, keeping them slightly raised, so that the beautiful black, white-centered spots on the forewings are visible, he opens and closes the front parts of his wings rhythmically, and

♂   ♀

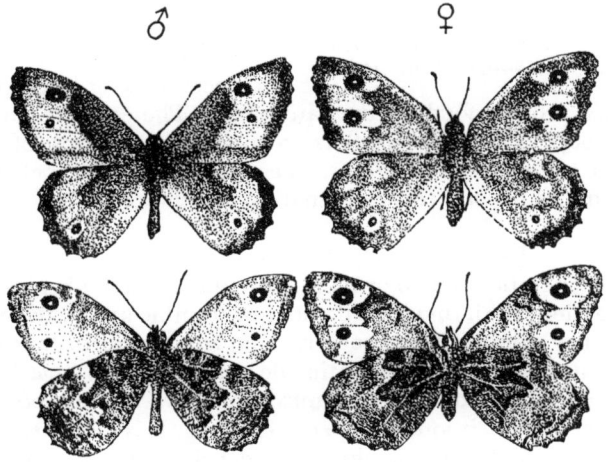

FIG. 8.—The Grayling. *Upper:* dorsal view; *lower:* ventral view. Scent scales on left wing of male outlined in black (*after Tinbergen et al., 1942*)

waves his antennae. This may last for several seconds, or even a minute. Then the forewings are raised and opened widely, and with a slow, quivering motion they are lifted so far that, although the male's body scarcely moves, it seems as if he is executing a deep bow in front of the female (Fig. 9). Then, still in this attitude, he folds the two forewings together, capturing the female's antennae between them as he does so. This whole bow takes somewhat more than a second. He then withdraws his wings, and quickly walks round the female until he stands right behind her, when he brings his abdomen forward and

makes contact with her copulatory organs. When the male succeeds in doing this, he turns round until he faces away from the
female, and in this position copulation is performed. After

about thirty to forty-five
minutes contact is broken
off, and the two individuals
leave each other for ever.
For the rest of its life, the
Grayling is an individualist
and never really associates
with others. The female lays
her eggs on carefully selected
objects among the grasses

FIG. 9.—The 'bow' of the Grayling
(after Tinbergen et al., 1942)

which provide food for the caterpillars. The eggs are not laid
in groups, and the caterpillars do not associate as those of many
other species do. There is, therefore, no social behaviour
extending beyond the short mating association.

TYPES OF SOCIAL CO-OPERATION

Co-operation of two or more individuals usually begins with
attraction; the individuals do not just stumble upon each other,
but they approach each other, often from great distances. In
the course of April the Nightingale males (*Luscinia megarhyncha*)
arrive in their breeding haunts. Their arrival can be easily
detected by their loud and persistent song. It is quite attractive
work to watch them in the early morning. It is soon obvious
that each male confines its wanderings to a small piece of
ground, its territory. Also, the males are all solitary: there are
no females yet. If we watch a male day after day, we will one
day discover that a female has arrived and joined the male.
From now on they form a pair. Knowing that such a female
has travelled alone, and many days later than the male, over
the whole distance from its winter quarters round the Mediterranean to our latitude, one realizes how remarkable it is that
it finds a male. How does it manage to do it?

Another, equally astonishing example is supplied by the
Emperor Moths, *Saturnia*. A southern species, *S. pyri*, has been
studied by the famous French entomologist, Fabre. He reports

how a female, hatched from the pupa in captivity, was surrounded soon after hatching by numerous males, some of which must have come from a considerable distance, since the species was rare in that region. Similar observations can be made in numerous other species of moths, such as the Lasiocampidae, the Lymantriidae, and the Psychidae.

Such instances strike us because they show achievements which at first glance seem to be much better than our own. Essentially however they are no more mysterious than are the achievements of so many other species in which the individuals come together from much shorter distances. Mosquitoes in a swarm, Porpoises in the sea, a Goose family in the farmyard, and innumerable other examples of animals coming or staying together are just as puzzling as the two cases mentioned above. First of all, we do not know what sense organs are involved. Do they see each other? Or do they hear, or smell each other? Or do they use sense organs unknown to us? And if we know what sense organs are used, why do they follow the message conveyed by them? How do they 'know' what they mean? In short, what is the mechanism of co-operation? Further, we want to know what purpose their aggregating serves. In the case of mating, we know the purpose. But what is the use of Starlings or Swallows flocking together? Or, concentrating our attention on a detail: what is the function of the male Grayling's bow?

Once the animals have come together, we see co-operation of numerous kinds. The simplest kind of co-operation is 'doing the same thing' as others. When one Herring Gull flees, the others flee as well. When Domestic Hens, even after they have just satisfied their hunger, see one of their number beginning to eat again, they will join it and will all start anew, just as the Sticklebacks already mentioned.[37] This principle of 'sympathetic induction', as McDougall has called it,[58] can be seen at work in many social animals, Man included. We yawn when we see another yawn, we get scared when we see signs of intense fear in another man. It has nothing to do with imitation; the reacting individuals do not learn to perform certain movements by watching others perform them, but they are brought into the same mood, and react by making their own innate movements.

Watching the flight manœuvres of a flock of Starlings or waders reveals another type of co-operating. Such animals not only fly when the others fly, but they direct their flight to that of the others. It is highly fascinating to see how thousands of Starlings, flying round above their roost on a winter evening, turn as if at a command; left, right, up, and down. Their co-operation seems so perfect that one forgets the individuals and automatically thinks of them as one cloud, as one huge 'super-individual'.

In all these cases the animals participating do the same thing. In many other types of co-operation, however, there is division of labour. In birds of prey, for instance, the male usually does the hunting for the whole family, while the female guards the brood. The male brings the food to the nest but it does not itself feed the young; it hands the prey to its mate (Fig. 10) who then

FIG. 10.—Male Kestrel passing prey to female.

feeds it to the young. Many birds begin a second brood before the first brood are capable of taking care of themselves. This means that the parents must incubate the new eggs and guard the young of the first brood at the same time. In the Nightjar (*Caprimulgus europaeus* [44]) these duties are divided; the male stays with the young and the female sits on the new clutch. In the Ringed Plover (*Charadrius hiaticula*), male and female take turns, and now and then relieve each other, the partner that had been guiding the young going to the nest, and the sitting bird walking away to the young.[49] This of course requires close co-operation and synchronization.

Division of labour is carried to an extreme in Honey Bee communities. The queen alone lays the eggs. The males have no other duty than to fertilize the virgin queens. All other duties are performed by the workers, infertile females. Some of them build the combs, others feed the larvae, others guard the hive and drive off intruders, others again fly out and collect honey or pollen, &c.

The division of labour is often reciprocal. The mating behaviour of the Grayling, and of course of many other species, are good examples. The courtship of the male stimulates the female to co-operate, and in actual coition the movements of male and female fit perfectly together, as do the copulatory organs involved. Innumerable are the ways in which this co-operation is effected in various species. Often it is still more complicated than in any of the species I have mentioned: think of dragonflies, squids, snails, or newts. However, even the simplest case of reciprocal co-operation offers abundant unsolved problems. Anybody who is willing to spend an hour in watching a Blackbird or other songbird feed its young can see such co-operation in action. All the time the parents are away foraging, the young are lying quietly in the nest. As soon, however, as the parent alights on the edge of the nest, the young get up, stretch their necks, and 'gape' (Fig. 11). The parent reacts to this by bending down, and depositing the food into the mouth of one of them. The young one swallows the food, and then subsides again. This is not the end, however: usually the parent waits and looks down into the

FIG. 11.—Blackbird feeding young

nest with close attention. Soon we see a movement in the cluster of young: one or two of them begin to waggle their abdomens; a circlet of spiny feathers round the cloaca is spread, and all at once a neat white package of faeces appears through the cloaca. It is picked up by the parent, who swallows it at once, or carries it away and drops it some distance from the nest. In this way nest sanitation is effected by co-operation: the young deftly presenting its load to the parent, the latter picking it up and disposing of it.

Mouth breeding fish offer another example of reciprocal behaviour. The female of the large Cichlid, *Tilapia natalensis*, for instance, picks up the eggs immediately after the male has fertilized them, and carries them in her mouth. When the young hatch, they remain in the mouth at first, but after a few days they swarm out, remaining however in the mother's neighbourhood. In case of danger the young swarm back to the mother's mouth (Fig. 12), and then she picks them up and keeps them until the alarm is over.

FIG. 12.—Young *Tilapia natalensis* returning to female

These short stories of co-operation between individuals could of course be multiplied. The spectrum of phenomena found in the animal kingdom is much more varied than I can record in the limited scope of this book.

On the whole, all the known phenomena can be grouped under four headings, which are each represented in one or more of the species treated here.

First, male and female come together for mating. Their co-operation leads to fertilization and the growth of new individuals. It serves an end which cannot be attained by either one of the partners alone. Usually, male and female both take an active part, although the male is usually more active than the female.

Second, parents, or one parent, guard or take care of the offspring as long as they are dependent on this care. Here the relationship is one-sided as regards effect. The parents 'help' the young, but the young do not 'help' the parents. Yet, as a first superficial observation suggested, and as analysis will confirm, there is mutual co-operation in so far as the young stimulate the parents, releasing their responses, just as much as the parents stimulate the young and release their responses.

PLATE 1

Male Three-spined Stickleback, adopting the threat posture in front of its
reflection in a mirror

Third, the association between individuals extends, in many species, beyond family life into group life. Since this group life shows so many aspects found in family life as well—it is even probable that in many species group life is nothing but an extension of family life—family and group organization will be discussed together.

Finally, individuals may associate in quite a different way: they may fight. At first sight fighting may seem to be exactly the opposite of co-operation; it is antagonism. Yet, as I hope to show, fighting between animals of the same species, although not of use to the individuals, is highly useful to the species, however paradoxical this may sound. The dangers to the individual are not essentially different, though different in degree, from the dangers which the individual encounters in mating and in guarding or defending the young. It is however obvious that mating and defending the young serves the offspring, and through it the species, whereas this is not immediately obvious in fighting. We will see in Chapter IV that fighting serves a function, and therefore we will have to analyse this type of co-operation as well.

The next chapters will be concerned with an investigation of the manifold ways in which various species manage to organize these four types of social co-operation; they will be arranged according to the functions they serve, and not according to the underlying mechanisms.

3

CHAPTER II

MATING BEHAVIOUR

THE FUNCTIONS OF MATING BEHAVIOUR

MANY animals, particularly species living in the sea, ensure the fertilization of the egg cells in such a simple way that we can scarcely speak of mating behaviour. Oysters for instance simply eject their sperm cells in huge numbers at a certain time of the year; for a while each individual is enveloped in a cloud of sperm cells. The egg cells, it seems, cannot avoid being fertilized. Yet even here an important sort of behaviour is involved: fertilization would not succeed if the various oyster individuals did not produce their sperm cells and their eggs at the same time. A certain synchronization therefore is necessary. As I hope to show, this applies equally to land animals.

In many higher animals, particularly land animals, fertilization involves mating, or copulation. This requires more than mere synchronization. It means bodily contact. This is a thing most animals avoid. This avoidance is an adaptation, part of their defence against predators. Being touched usually means being captured. Also, during actual mating the animals, and above all the females, are in a dangerous, defenceless position. In such animals the mating behaviour therefore involves the suppression of the escape behaviour. Since the female carries the eggs for some time, often even after fertilization, and since in so many species the female takes a larger share than the male in feeding and protecting the young, she is the more valuable part of the species' capital. Also, one male can often fertilize more than one female, an additional reason why individual males are biologically less valuable than females. It is therefore not surprising that the female needs persuasion more than the male, and this may be the main reason why courtship is so often the concern of the male. Often the male needs persuasion as well, but for a different reason. The males of most species are extremely pugnacious in the mating season, and unless the females can appease the males, they may be attacked instead of courted.

Further, apart from synchronization, which is a matter of co-ordinating the time pattern of mating, there must be close spatial co-ordination: the males and females must find each other; during actual copulation they must bring their genital organs in contact with each other; and, next, the sperm must find the egg cell. This orientation is also a task of mating behaviour.

Finally, there is a premium on the avoidance of mating with members of another species. Since the genes, and the highly complicated growth processes started by them, are different in each species, mating between animals of different species brings widely different genes together, and this easily disturbs the delicately balanced growth pattern. Mating between different species therefore often results in fertilized eggs which are unable to live, and which die at the beginning of their growth; in less serious cases the hybrids may live but are less vital, or infertile. This premium on intraspecific mating has led to the development of differences between the mating patterns of different species, so that each individual can easily 'recognize' its own species.

Apart from actual insemination, therefore, synchronization, persuasion, orientation, and reproductive isolation are the functions of mating behaviour.

Our problem in this chapter is: how are these functions fulfilled? What part does social behaviour play, and how does it attain these results? Let me say right at the beginning that our knowledge is very patchy. We have bits of information on each of these problems, but part of our knowledge applies to one species, and other parts to other species. In not a single species do we know the whole picture. The only thing I can do, therefore, is to present some examples of the various ways in which mating behaviour attains these ends, leaving it to future research to find out to what extent we are entitled to generalize our findings.

One thing seems to be obvious already: all the behaviour involved is of a relatively low 'psychological' level, and does not imply foresight of these ends, nor deliberate action with the aim of attaining them. As we shall see, mating behaviour in all animals except Man and, perhaps, some of the apes, consists of immediate reactions to internal and external stimuli. There is no way in which 'foreseen' effects of the behaviour can be brought into play as causes of the behaviour, as it does, in some as yet completely mysterious way, in Man.

SOME INSTANCES OF TIMING

The timing of the reproductive behaviour of Oysters (*Ostrea edulis*) has been shown recently [41] to be the work of a rather unexpected outside factor, and therefore it is not, strictly speaking, a sociological problem. Yet it is perhaps useful to discuss it here as an example of the way in which the action of outside factors may often, so to speak, 'fake' social co-operation.

About eight days after the oysters spawn, the larvae 'swarm'. They lead a very short floating life, and soon settle down on a solid substratum. In the muddy estuaries of the Scheldt, in Holland, oyster breeders increase their oyster stock by depositing roof tiles as artificial substrates on the bottom of the sea. This must not be done too much in advance of swarming, since the tiles would then become overgrown with other organisms before the oyster larvae could settle. A zoologist therefore had to find out whether he could forecast when the swarming would take place. His forecast, based upon many years of study, seems astonishing: 'The big maximum in swarming is to be expected each year between June 26 and July 10, at about 10 days after full or new moon' (Fig. 13). This sounds like a fable, yet it is the hard truth. Since swarming takes place eight days after spawning, this means that spawning is to be expected two days after full or new moon. This gives us the key to the factor that is responsible for the timing: the tides. Spawning takes place at spring tide. How the spring tide affects the oysters is not yet known; it is not improbable that it is a matter of water pressure, which reaches its greatest oscillations at spring tide. Also, the intensity of the light penetrating to the bottom shows its maximum fluctuations at that time, and this might also be a factor.

Since the oysters do not spawn at each spring tide, there must be another factor preparing them to be ready in June to react to the spring tides; the nature of this factor is not known yet. It works much less precisely than the tides, for although the maximum of spawning occurs between June 18 and July 2, there are minor peaks during the preceding and the following spring tide. This factor is not known in the oyster, but in other animals we know something about it.

Not only the oyster but several other marine animals are known to be timed by the tides, among them the famous Palolo worm of the Pacific, and various other worms and molluscs.

Timing in the higher animals is a more complicated affair. Something is known of fish, birds, and mammals of the Northern temperate zone. Reproduction of most of these begins in spring. The first phase is migration towards the breeding grounds. This is done by all individuals at approx-

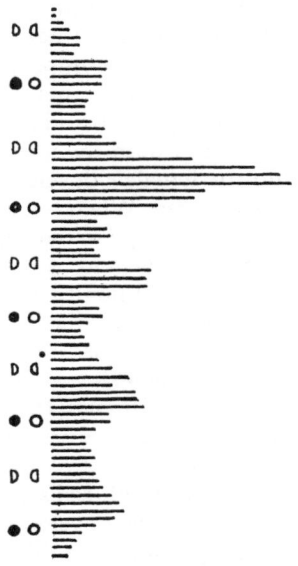

FIG. 13.—Swarming of oyster larvae on 74 consecutive days in June, July and August, demonstrating correlation with the phases of the moon (*after Korringa, 1947*)

imately the same time, though there may be weeks between the arrival of the first and the last comers. This rough timing is again due not to social behaviour but to reactions to an outside factor. The main factor here is the gradual lengthening of the day in late winter.[9, 76] Various mammals, birds, and fish have been subjected to artificial day-lengthening. The result was that the pituitary gland in the brain began to secrete a hormone which in its turn affected the growth of the sex glands. These then began to secrete sex hormones, and the action of these sex hormones on the central nervous system brought about the first reproductive behaviour pattern, migration. Often a rise in the temperature of the environment has an additional effect.

As I said, this timing process is not very accurate. The different individuals do not all react to the lengthening of the day with the same promptness. There may be a considerable difference between the male and the female of a pair. It has been found, in pigeons and in other animals, that if the male is further advanced than the female, his persistent courtship may speed up the female's development. This has been found in the following way. When a male and a female are kept separately in adjoining cages so that they can see and even touch each other, but are prevented from copulating, the persistent courtship of the male will finally induce the female to lay eggs.[14, 15] These of course are infertile. It may occur in captivity, when no males are available, that two female pigeons form a pair. Of these two, one then shows all the behaviour normally shown by the male. And although their reproductive rhythms may have been out of step at the beginning, the final result is that they both lay eggs at the same time. Somehow their mutual behaviour must have produced synchronization, not merely of behaviour, but also of the development of eggs in the ovary.

It is possible that this effect may be found in other species as well. It has been suggested by Darling[18] that the communal courtship of birds breeding in colonies may have the same effect.

A further refinement of synchronization however is necessary. In all species that copulate, and in many other species as well, the co-operation between male and female must run according to an exact time schedule, and without exact co-operation no fertilization would be possible. In only very few species can the male force the female against her will to copulate. This means that in many species some form of very accurate synchronization must occur, which is a matter of fractions of a second. This is done by a kind of signal system. As an example I will discuss the mating of the Three-spined Stickleback.[101] In the scheme of the mating behaviour (p. 12) the arrows indicate not merely a temporal sequence, but also a causal relation: each reaction really acts as a signal which releases the next reaction in the partner. Thus the male's zigzag dance releases approach in the female. Her approach in its turn releases leading in the male. His leading stimulates her to follow, and so on. This can easily be shown by the use of models or dummies. When a very crude imitation of a pregnant female is presented to the male in its territory (Fig. 14), he will approach it and perform the zigzag dance. As soon as the model is then turned in his direction and 'swims' towards him, he turns round, and leads it to the nest.

Pregnant females can be induced in a similar way to react to

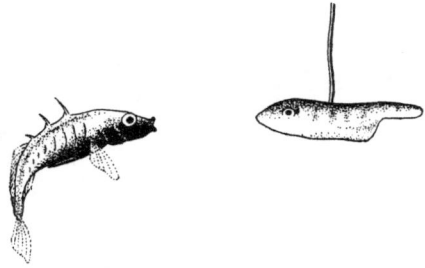

FIG. 14.—Male Three-spined Stickleback courting a crude model of female

a model of the male. A crude fish model again is sufficient, provided it is painted red underneath. A bright blue eye will also help, but beyond that no details are necessary. If such a model is moved round a pregnant female in a crude imitation of the zigzag dance, the female will turn towards the model and approach it. If we then let the model swim away, the female follows it, and it is even possible to make her try to 'enter' anywhere in the bottom of the aquarium by making the male model 'show the nest entrance' (Fig. 15). No nest is necessary; the movement of the model is sufficient stimulus for the female to react.

Now in these cases the fish do not react exclusively to the partner's movements, but also to certain aspects of shape and colour. If the female dummy does not have the swollen abdo-

men of the real female, it will not or scarcely stimulate the male
to dance. If the model of the male does not show a red underside, the female shows no interest in it. On the other hand, all
other details have little or no influence, so that it is easier to
release mating behaviour by using a very crude but 'pregnant'
dummy than with a live but non-pregnant female. However,
the swollen abdomen, and the red colour, which are displayed
continuously, are not responsible for the timing of the response
of the male. It is the movements, which appear suddenly and
immediately elicit a response, that are responsible for the exact
timing.

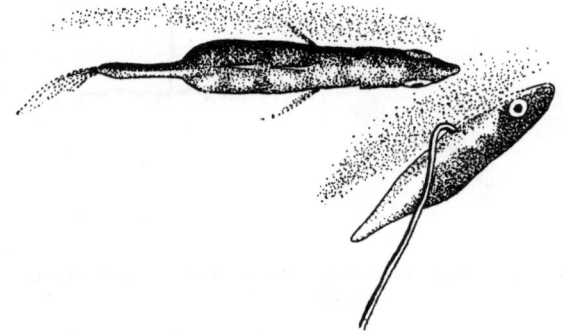

FIG. 15.—Female Three-spined Stickleback following male model
which 'shows the nest entrance', seen from above

The mating behaviour of Sticklebacks is a complicated series
of such signal-response sequences, and the end result is that the
male fertilizes the eggs immediately after they are laid by the
female. It is not at all difficult to observe this behaviour, and
to carry out all the model tests described. The Three-spined
Stickleback will readily breed in an aquarium of a cubic foot
or larger. It should have sand on the bottom, and plenty of
green vegetation, including some green thread-algae.

The mating behaviour of many species involves such signal-
movements which serve this ultimate refinement of synchronization.

PERSUASION AND APPEASEMENT

Even when an animal is in a sexually active condition, it does not always react immediately to the partner's courtship. It may take a considerable time to overcome the female's reluctance. The zigzag dance of a male Stickleback for instance does not always elicit the female's response at once. She may approach in a half-hearted way, and stop when the male tries to lead her to the nest. In that case the male returns, and again performs his zigzag dance. After a number of repetitions the female may eventually yield, follow him, and enter the nest.

A similar repetition of signals is necessary when the female has entered the nest. The male's prolonged 'quivering' is required to make her spawn. When you take the male away just after the female has entered the nest, she is unable to spawn. When you touch such a female gently with a light glass rod, imitating the male's quivering with it, she spawns just as easily as when the male has delivered the stimuli. Both male and rod have to touch her a great number of times.

In many species this repetition of signals is the rule. The copulation of Avocets for instance [60] is preceded by curious antics: both male and female stand and preen their feathers in a hasty, 'nervous' fashion. After some time the female stops preening, and adopts a flat attitude (Fig. 16). This is the signal indicating that she is willing to mate,

FIG. 16.—Pre-coition display of European Avocets (*after Makkink, 1936*)

and only then does the male mount and copulate. Sometimes he does not react at once, but only after a certain time.

Herring Gulls have a similar introduction to coition. Both male and female bob their heads upwards, uttering a soft, melodious call with each bob (Fig. 17). Here it is the male which takes the initiative in copulation: after a series of such mutual head-tossings he suddenly mounts and mates.

FIG. 17.—Pre-coition display of Herring Gulls (*after Tinbergen, 1940*)

Sometimes persuasion has another function. In many birds, and in other species as well, the males become very aggressive in the breeding season. Actually, most of the fighting seen in animals is fighting between rival males in spring. This fighting is essential. Since it is always aimed at a rival male, the female has to differ from the male lest she should be attacked as well. In species such as the Chaffinch, the Redstart, or various Pheasants, the differences in plumage partly serve this purpose. In many other species however, such as the Wren, the plumages of male and female are not very different, or they are even identical, and here the female has to show special behaviour to suppress the male's aggressiveness. The essence of this 'female courtship' therefore is to avoid provoking attack. Whereas a strange male may either flee from the displaying male—in which case it immediately elicits pursuit—or strut and threaten in reply (which also provokes the displaying male's aggressiveness), a female does neither. In the Bitterling (*Rhodeus amarus*) the female is at first attacked.[8] She either withdraws quietly or merely avoids the attack by swimming under the male. The male then seems to be unable to attack her, and after a while it ceases to try and begins to court the female (Fig. 18). A similar unobtrusive appeasement can be observed in many Cichlids.[5] In other species the female shows infantile behaviour, that is to say, it resorts to the same method of appeasement as employed by the young, which probably stimulates the male's parental drive. That is why in so many species the male feeds the female during courtship. This happens, as we have seen, in the Herring Gull. There are also species in which the appeasing postures used during courtship are different from those used by the young. The female, or in other species both sexes, then show a type of behaviour which in many respects is the exact

PLATE 2

The forward threat posture of the Black-headed Gull

Head-flagging of Black-headed Gulls

opposite of the threat behaviour. When, for instance, Black-headed Gulls (*Larus ridibundus*) meet in the mating season, they show the 'forward display', lowering the head and pointing the beak towards each other (Plate 2, top). This threat gesture is emphasized by the brown face, which surrounds the bill, the actual weapon. Mates however show their friendly intentions

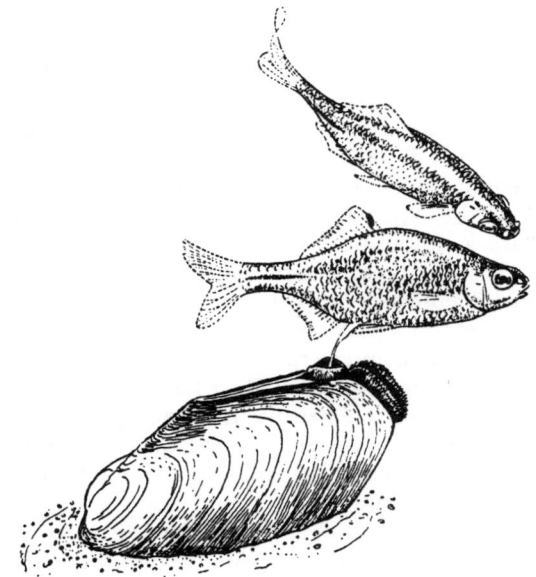

FIG. 18.—Male Bitterling courting the female during spawning
(*after Boeseman et al., 1938*)

by 'head flagging' (Plate 2, lower fig.); they stretch the neck, and then, by a sudden jerky movement, they turn their faces away from each other.[109] Here, since both sexes are rather aggressive, the male appeases the female, as well as vice versa.

In some web-building spiders, the male visits the female on her web. Here the male has to appease the female, because he might be mistaken for prey.

The spatial directing of mating movements is another important function of courtship. The most obvious function to be fulfilled is attraction. Many songbirds, such as the Nightingale, spend the winter far from the breeding grounds. The males, as mentioned above, return from the south well in advance of the females. How do the females find the males? This is made possible by the song. Many birds attract the other sex by some loud noise. In the Nightingale we happen to find this noise beautiful and have called it song. But the spring call of a male Grey Heron (*Ardea cinerea*), a harsh cry, does not appeal to human ears. Yet it does to the female Heron.[113] It serves exactly the same function as the Nightingale's song. The Nightjar's rattling, the Woodpecker's drumming and the croaking of toads (Plate 3, lower fig.) belong to the same category. So exactly is the song of many birds tuned to this function that the song is most intense as long as the males are still unmated, and it stops as soon as a female arrives. This again is due to a conflict between various interests. Song serves the species in that it attracts females (and, as we shall see later, repels rival males), but it endangers the male because it attracts predators as well. Nature, as always, has evolved a compromise: song is only produced when it is really needed, or at least when the advantages outweigh the disadvantages.

Since most animals are deaf (only the vertebrates and some other groups are exceptions), we find auditory advertisement in relatively few groups. It is well developed in birds, in frogs and toads, and in various insects such as crickets and grasshoppers. Special organs have evolved in such groups exclusively for the production of sound.

Other groups use scent as a means of attracting the opposite sex. Extreme cases are found among moths. The Psychid Moths [62] have been studied to some extent and will be chosen as examples. The females have lost the capacity to fly; they are practically wingless. Soon after hatching, a female leaves the tubular shelter in which she has been living as a caterpillar and as a pupa. She does not move beyond her doorstep however, but remains hanging beneath her shelter. The males can fly. Shortly after hatching they leave their house, and take wing in order to search for a female. This search is guided by a scent which emanates from a virgin female. This attraction by female scent is highly developed in many other moths, such as Saturnia (Fig. 19) and Lasiocampa species. In such species the male is often able to find a female from a considerable distance, and his organs of smell, situated on the plume-like antennae, are highly sensitive. It is not at all difficult to collect caterpillars of those species, let them pupate and hatch, and watch the wild males come and enter the house in their search for the virgin females.

FIG. 19.—*Saturnia pyri*. The olfactory organs on the antennae are strongly developed in the male

Visual attraction plays a part in many species. It is beautifully developed in the Sticklebacks. The male Three-spined Stickleback develops its most brilliant nuptial colours after the nest is finished. The red of the underside becomes more brilliant, and the dark shade which has covered his back during nest building becomes a fluorescent bluish white. Simultaneously, his behaviour changes. While during nest building he moved about smoothly, avoiding sudden movements, he now keeps swimming round the territory in a jerky abrupt fashion, which together with his conspicuous dress makes him visible from afar.

Many birds add visual displays to their auditory attraction devices. This is developed most impressively in birds of wide open plains. The waders of the Arctic tundra and many marsh birds in this country specialize in this respect (Fig. 20). Again

we often find a combination of conspicuous colours and movements. Lap-wing, Black-tailed Godwit, Dunlin and other waders are good examples. Other species have entirely specialized on movement, and are lacking in colour; such are found among the more vulnerable songbirds: Pipits, Larks. Specialization on colour has also occurred: the Ruff (*Philomachus pugnax*) has no special song flight, but relies on gorgeous coloration. Yet it has evolved another signal movement: now and then the males on a 'lek' lift their wings, the light undersides of which make them very conspicuous (Plate 4, upper fig.). This wing lifting occurs particularly as a reaction to females flying in the distance, and it seems to attract these females. These lek birds apply still another principle, which has been called the 'flower-bed principle': by crowding together their individual colour-effects are added together; they form a large gaudy patch somewhat like a flower bed.

FIG. 20.—Lap-wing in flight

In only a few of these cases has experiment proved the attracting influence. The red colour of male Sticklebacks has been proved to attract the females; models lacking red do not attract them. The influence of song has been nicely demonstrated in various locusts; Fig. 21 illustrates such a test. In one cage, hidden in the heather, singing males of Ephippiger were kept; in the next cage there were the same number of males, which were silenced by gluing their stridulation organs together. This is a minor operation, which leaves this wingless form free to pursue all other activities. At a distance of ten yards, females in mating condition were released. They invariably made their way in a short time to the cage containing singing males.[20]

Experiments of this type justify the conclusions drawn in these paragraphs concerning the attracting influence of various types of display. Yet further experimental work is needed.

The orientation task of courtship is not finished when

attraction has been effected. In actual copulation, the male has to bring his copulation apparatus into contact with that of the female, and this again requires powers of orientation. This is most obvious in many insects, where the males possess a complicated system of claspers to be fitted into the closely corre-

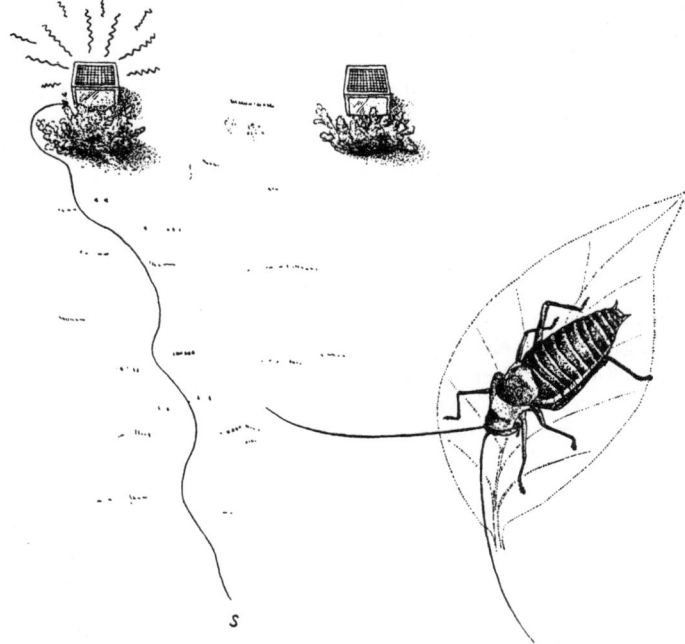

FIG. 21.—An experiment on the function of song in the locust *Ephippiger* (*after Duym and Van Oyen, 1948*)

sponding 'negative' counterparts of the females. But in less 'mechanized' animals such as birds this problem exists as well: the male cannot bring its cloaca in touch with that of the female without first reacting to orienting stimuli from the female. However, very little is known about these behaviour mechanisms.

4

REPRODUCTIVE ISOLATION

Hybridization between species is exceedingly rare in nature. This is only partly due to differences in habitat preference between different species. Closely related species which breed in entirely separated geographical regions, and species which, though living in the same general region, go to different habitats to breed, are prevented from cross breeding by this spatial separation. But even when there is no such separation, species do not ordinarily interbreed. This is due to the fact that the various signals serving attraction, persuasion, appeasement, and synchronization, are so very different from one species to another. Also, the tendency to react to such signals is specific; every animal is innately equipped with the tendency both to give its species' signals and to react only to the signals of its own species. Yet one often sees, in nature, sexual reactions to other species. The males of the Grayling, which I have studied for several seasons, start their courtship by following the females in flight. This sexual pursuit is not released by females only: butterflies of other species, beetles, flies, small birds, falling leaves, even their own shadow on the ground attracts them. How is it that they never mate with these other species? Similar observations, leading to the same query, can be made on birds, fish, and many other animals.

The answer seems to be found in the chain character of mating and pair formation activities. When a female Grayling is willing to mate, it reacts in a special way to the male's sexual pursuit: it alights. All the other species usually do the opposite: if bothered by a pursuing male, they fly off as fast as they can, and this makes the male abandon its pursuit. Only closely related species react occasionally (Plate 3, upper fig.) but this has never been observed to lead to mating. Sticklebacks show essentially similar behaviour. The male may react to a small Tench entering its territory by zigzag-dancing. For a continuation of his mating behaviour, however, it is necessary that the partner swims towards him. Even if a Tench does this inadvertently, it has to follow the male to the nest, it has to enter the nest, and it has to spawn there before it can release sperm-ejaculation

PLATE 3

A male Grayling (left) courting a female of the related species
Hipparchia statylinus

Singing male Natterjack

in the Stickleback. In other words, it must show the correct
series of responses to the whole succession of the male's courtship activities, including the final 'quivering'. And this is so
extremely improbable that it has never been observed. The
sign stimuli of each separate reaction of the chain may not be
sufficient to prevent reactions to other species, but since the
separate reactions are each released by different stimuli, these
together are sufficiently typical to prevent interspecific mating.
This is obvious in species with 'mutual' courtship, for here
each sex shows a series of courtship activities. But even in a
species such as the Grayling, where the female just sits while
the male performs his courtship sequence, the female supplies
a series of stimuli: experimental analysis has shown that the
various activities of the male, as described in Chapter I, are
released by stimuli which differ from one reaction to the next.

This specificity is particularly needed in closely related
species. As we will see later, the behaviour patterns of closely
related species are always very similar, just as their morphological characters are. They simply have not had the time for
wide evolutionary divergence. But in such species there is
always some striking difference between mating patterns, at
least if spatial (geographical or ecological) or temporal (differences in breeding season) separation does not render this un-
necessary. For instance, the Ten-spined Stickleback's (*Pungitius
pungitius*) mating behaviour is rather similar to that of the
Three-spined species.[83] It has however evolved very different
nuptial colours in the male. The male of the Ten-spined species
is pitchblack in spring (Fig. 22). Just as the red colour attracts
Three-spined females, so the black colour appeals to the females
of the Ten-spined species. This, together with some minor behaviour differences, is sufficient to make interbreeding rare.

A systematic study of this problem of reproductive isolation
has been undertaken in only one group: the fruit flies (*Drosophila*).[84] The first results indicate that mating attempts between different species break off at various stages of the courtship, depending on which species are involved. Whenever such
an interruption of courtship is found to be consistent in a series
of observations, it is a sign that we are dealing with a specific

response which cannot be released by the partner. The results obtained thus far show that in some cases the male fails to give the correct stimulus, in other cases the 'fault' is with the female.

FIG. 22.—Male Ten-spined Stickleback showing nest entrance to female (*after Sevenster, 1949*)

CONCLUSION

This very brief and sketchy review may be sufficient to show the intricate nature of behaviour patterns serving co-operation between the members of a pair. It has been shown that we must distinguish between four different types of functions served by courtship. This does not mean that each particular courtship activity serves only one of these ends. The zigzag dance of the male Stickleback for instance certainly serves timing, persuasion, orientation, and isolation, but the difference between the nuptial colours of Three-spined and Ten-spined Sticklebacks can only be understood from the viewpoint of isolation. Also, we know of courtship activities which have to do with timing and persuasion but not with orientation: Grayling females for instance can be timed and persuaded by the courtship of one male, and then mate with another male, which shows that the first male did not orient the female's response towards himself.

Similarly, in pigeons, the persistent cooing and bowing of the male does not so much orient the female but it makes her gonads start ovulation. The various closely related 'Darwin's Finches' of the Galapagos islands were found to have almost identical courtships.[48] Yet there is no interbreeding. Here reproductive isolation is effected partly by ecological isolation, partly by each species reacting specifically to its own species' type of bill, which, in relation to the type of food taken, is different from one species to another. In this case therefore the courtship activities have nothing to do with reproductive isolation, but they do serve all the other functions.

In all these cases courtship activities, however different their functions may be in detail, have one thing in common: they send out signals to which the sex partner responds. In a later chapter I will discuss the nature and function of these signals more closely. It will then become clear that many of the conclusions and generalizations are still tentative, because experimental evidence is fragmentary. Further experiments with the aid of models are much needed.

CHAPTER III

FAMILY AND GROUP LIFE

INTRODUCTION

WHEREAS in Chapter II we dealt with relationships between two partners co-operating to achieve one end, co-operation in the family is more complicated, for it involves relationships between male and female, and relationships between parents and offspring as well. Also, the ends at which the activities are aiming are really more complicated. The parents have to provide shelter and food, and they have to defend the young against predators. In all these functions, the activities must be timed and oriented. Some other tendencies which would otherwise interfere with this have to be suppressed: in many species, for instance, the young provide all the stimuli normally releasing eating in the parents. In others the parents provide all the stimuli necessary to release escape in the young. Further, there is need for reproductive isolation, or the prevention of reactions to the young, or parents, respectively, belonging to other species; for such reactions would be a loss of efficiency, and inefficiency means defeat in the struggle for existence. Further, a new element enters into the situation, which was absent, or at least not outstanding in the relationships between mates: defence of the brood against predators. This is one means of compensating for the helplessness of the young.

However, neither the suppression of other drives nor the prevention of interspecific co-operation are up against such powerful odds as in mating behaviour, and this is probably the reason why co-operation in the family does not depend on such elaborate ceremonies as are found in courtship. Since every 'ceremony', or use of signals, renders the performing individual conspicuous and therefore vulnerable, such ceremonies have only been allowed to evolve when the advantages outweighed

PLATE 4

'Ruffs' on 'lek'. The male on left displays the white underside of the wings as a reaction to a female in the distance

Unusual nest relief by a Lesser Black-backed Gull. The relieving bird (right) tries to push its mate off the nest

PLATE 4

the disadvantages. In other words, they have not been evolved beyond what is strictly necessary. This might seem strange in view of the abundance of signals used in various species, but this doubt disappears when one realizes the strict necessity of signals. We tend to consider social co-operation, such as the raising of chicks by parent birds, as something quite common-place. This however is merely because we are used to it. Instead of being astonished when abnormal parents desert their chicks we should be astonished that most parents do not, and manage to bring this very difficult and complicated task to an end.

We will first consider the organization of the family, then that of the group, and in both cases we will have to examine the nature of the relationships that constitute this organization.

FAMILY LIFE

When a Herring Gull happens to find an egg on its territory before it has a clutch of its own, it will not brood that egg, even if it is found right in its nest. This is not because it sees that the egg is not its own, for it does not normally distinguish between its own eggs and its neighbour's. It is because the gull is not yet 'broody', that is, certain internal factors without which the incubation response is impossible, are still absent. Outside the breeding season, an egg is just food to a gull. Shortly before the eggs are laid, both the female and her mate undergo an internal change by which their nervous system becomes ready to respond to the stimulus-situation 'eggs in nest' by brooding. The main internal factor responsible is, in pigeons and fowl, and most probably also in the gull, a hormone secreted by the pituitary gland, prolactin.[73] Yet this is not the only factor responsible for the timing of incubation, for incubation on the empty nest, although it occurs, is not the rule, and it is never very persistent. The eggs themselves are necessary; they provide visual and tactile stimuli which in a broody bird release sitting. Here again therefore we have to do with a timing in stages: a rough timing by the hormonal situation, and a more detailed timing by stimuli evoking an immediate response.

When the chicks hatch, the behaviour of the parents changes again. New patterns such as feeding them, and guiding them, appear. These new patterns are again different from one species to another, and it would be worth while to give a review of the innumerable types of parental behaviour found in nature. The scope of this book however does not permit this. I can merely stress the need of further (even of purely descriptive) study, since our knowledge is still very incomplete.

The change from care of the eggs to care of the young is again a matter of rough internal timing, corrected by a more accurate timing by external stimuli. At the beginning of the incubation period, for instance, a bird will not readily accept a pipped egg or a young bird. But towards the end such an egg, or a young

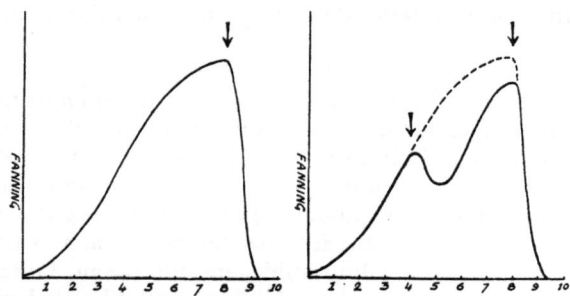

FIG. 23.—*Left:* Time spent fanning by male Three-spined Stickleback in the course of ten days from spawning. Arrow indicates hatching date. *Right:* Fanning graph when eggs are replaced on 4th day by hatching eggs. Second arrow indicates 'autonomous' fanning peak not induced by foster eggs

bird, will be accepted, even if it is offered some days early. In the course of incubation therefore the bird is being prepared internally for the next phase. The nature of this internal change is not known. It is certain that prolactin again is necessary, but since prolactin activates incubation as well, some additional change must take place.

The external stimuli are provided by the young. In the case of some birds, there are indications that the stimulus is given by chicks while still in the egg.[100] Most probably the parents react to the calls, which can be heard before hatching.

In the Stickleback, exchange of eggs against older eggs makes the male fish accept the young when they come; he guards them, and his fanning activity shows a sudden drop as soon as the foster young hatch. However, it does not stop entirely, and reaches a new, lower peak (Fig. 23) at the time when his own eggs would have hatched had he been allowed to keep them. Since the male is then no longer in touch with his own young, this second peak must have been caused by internal factors.

Once the parents have entered into the phase of care of the young, their activities such as feeding the young must be timed by the latter. Again, birds have been studied best in this respect.

In many song birds, the young have to gape in order to enable the parents to feed them. If they fail to gape, the parent bird looks at them, then looks around 'help-lessly', as if quite at a loss. It may resort to special social behaviour to stimulate them, such as touching them, or uttering soft calls, and if this fails, the parent usually swallows the food. Nature often provides us with an experiment. Most instructive, for instance, are the reactions of songbirds to a young Cuckoo (*Cuculus canorus*). When a Cuckoo has laid its egg into the nest of a Redstart (*Phoenicurus phoenicurus*), the Cuckoo's egg usually hatches some time in advance of the foster parents' eggs. Shortly after hatching, the young Cuckoo proceeds

FIG. 24. — Young Cuckoo throwing out an egg of its foster parents (*after Heinroth and Heinroth, 1928*)

to throw out the other eggs (Fig. 24), or, if they have hatched in the meantime, the young. It takes them on its back, and, crawling backwards, it pushes them over the nest's rim.[31] Such hapless nestlings die of cold and starvation. They may lie on the rim of the nest. It does not occur to the parents to save them by bringing them back into the nest or by brooding and feeding them where they are. And since the young Cuckoo gapes intensely, and the young Redstarts do not, the parents just ignore them; they do not receive the necessary stimuli. In several birds of prey it has been observed that the order in which the young are fed depends entirely on their begging.[80]

Those that beg most intensely receive the food. In this way each young one will usually get its turn, for the intensity of begging depends on how hungry they are, but when one of the young happens to be weak right from the beginning, as often happens in harriers, and in various owls, it may get too little food, its begging response becomes increasingly weaker, and it eventually dies.

Not only must the parents react to the children, but the children must react to the parents so as to time their begging behaviour. Begging again, just as any other type of advertisement, is dangerous, and continuous begging is a luxury which only few species can allow themselves.[96] It is found in some hole-breeding birds such as woodpeckers. Even there though it may lead to destruction. The Little Owl is known to rob nests of hole-breeding birds such as Starlings, and I have myself watched a Goshawk taking one noisy Black Woodpecker after the other from the nest hole. As a matter of fact, begging is often restricted to the short moments when one of the parents is actually at the nest with food. This again is possible because the young react to stimuli given by the parents. Young nestling thrushes for instance begin to gape when the nest is slightly jarred by the parent bird's alighting on it. Later, when their eyes have opened, they react to a visual stimulus. Also, some broods begin to react, after a week or so, to the voice of the parent. A Herring Gull chick is stimulated to beg by the parent's 'mew-call'; it then runs up to the parent and begins to peck at its bill tip, where, after some failures, it gets hold of the food presented, and swallows it. After a few days, the chicks learn to know their parents individually, and beg only from them. The chicks of domestic Fowl are more independent, and feed on their own right from the beginning. Yet they run to the mother, reacting to her special calls and movements whenever she has discovered something edible.

As will be clear from these few examples, the functions of timing and orienting are usually fulfilled by the same stimuli. The parents convey to the young not merely 'now there is food', but also 'here is food'.

Persuasion, or the suppression of inappropriate responses,

offers new and very interesting problems. In many fish, for instance, the young are so tiny that they fall well within the category of creatures that make up the food of their parents. This applies for instance to the Sticklebacks and to Cichlid fish, both groups which show extensive care of the young. How is it possible that the parents do not eat their own young? In mouth-breeding Cichlids this is solved in a relatively simple way. As long as the female *Tilapia natalensis* carries the young about in her mouth, she simply does not feed; the whole feeding instinct is suppressed and with it the tendency to eat the young.[5] But other Cichlids have the same tendency as the male Stickleback: they pick up stragglers and bring them back to the flock.[4] Their foraging instinct is not extinguished, and they eat *Daphnia* or *Tubifex* or other prey quite happily all through the breeding season. Lorenz reports a highly interesting and amusing incident which demonstrates the power of the parents to distinguish between food and their young.[57] Many Cichlids carry the young back, at dusk, to a kind of 'bedroom', a pit they have dug in the bottom. Once Lorenz watched, together with some of his students, a male collecting its young for this purpose. When it had just snapped up a young one, it eyed a particularly tempting little worm. It stopped, looked at the worm for several seconds, and seemed to hesitate. Then, after these seconds of 'hard thinking', it spat out the young, took up the worm and swallowed it, and then picked up its young one again and carried it home. The observers could not help applauding enthusiastically!

In many birds, the grown-up young, assuming the shape of the adult bird, begin to 'annoy' their parents, that is to say, they begin by their very shape to stimulate the aggressiveness of their parents. They can prevent attacks for some time by showing infantile behaviour that cannot be misunderstood by the parents. I have watched this in the Herring Gull, and found that the young develop a submissive attitude which is in a sense the opposite of the aggressive posture of an adult Herring Gull: they withdraw the neck, adopt a horizontal attitude and point the bill slightly upward (Plate 5, upper fig.). It is certainly no accident that this posture is identical with the posture adopted by the female when she approaches a male in order to 'propose' to him (Fig. 25). As the season proceeds, this posture of the young becomes less and less effective however, because the

FIG. 25.—Female Herring Gull (*left*) proposing to male

parental drive in the adults wanes. As usual, these relationships are balanced in such a way that, when the parents lose interest in the young, these are able to look after themselves.

Another system which prevents the parents from eating their own young has been developed in Cichlid fish. They do eat young of other species. A curious learning process makes them distinguish their own young from those of other species. This was shown in a simple experiment by Noble.[69] He replaced the eggs of an inexperienced pair, who were breeding for the first time in their lives, by eggs of another species. When the eggs hatched, they accepted the young, and raised them. Whenever they met young of their own species however they ate them! Such a pair was spoiled for ever, for next time, when they were allowed to keep their own eggs, they devoured the young when they hatched! They had learned to consider another type of young as their own.

Conversely, the young must somehow be protected from reacting to their parents as they would to predators. Young fish of many species escape from fish the size of their parents. This of course would make parental care impossible if escape from the parents themselves were not in some way prevented. In the Stickleback, I have the impression that the male is just too quick for the young, and overtakes them when they try to flee. It is quite an amusing sight to see the male trying to catch his offspring while they are doing their best to escape. As I described in Chapter I, when the young go to the surface for the

purpose of filling their swim-bladders, they develop an astonishing speed, specially reserved for this action. They then really
outwit their father and manage to dash up and down again
without giving him a chance to catch them, although he always
tries. Cichlid young completely trust their parents; how that is
brought about, I do not know.

Lorenz has discovered how Night Herons (*Nycticorax nycticorax*) manage to 'convince' their young of their parental intentions.[55] When a Night Heron comes
to the nest, it makes a perfect bow
towards the inhabitants, whether
these be its mate or its young. In
doing so, it shows off the beautiful
bluish-black cap, and raises the three
thin white plumes which, when at
rest, are folded together (Fig. 26).
After this introduction, it steps down
into the nest and is received cordially.
One day when Lorenz happened to
have climbed a nest tree and was
standing at the side of the nest, one
of the parents returned. Being a tame
bird it did not flee, but adopted an
aggressive posture instead of the appeasement ceremony. The young,
which were not afraid of Lorenz, at
once attacked their father. This was
the first indication, which he later
confirmed, that the young 'recognize'
their parents because these are the

FIG. 26.—Night Heron at
rest (*above*); performing
the 'appeasement cere-
mony' (*below*)

only Night Herons, in fact the only birds, which show the
appeasement ceremony, thus suppressing their defence.

Not much is known about this problem of the suppression of
inappropriate responses, but these few examples show how very
interesting a problem it is. It is not at all difficult to do some
research on it even in the field; when the problem is recognized
many types of behaviour become intelligible which seemed to
make no sense before. In the Cichlid *Hemichromis bimaculatus* for

5

instance, where the parents take turns in guarding the brood, the parent that has just been relieved by its mate swims away from the brood in a quick, straightforward manner (Fig. 27);

FIG. 27.—*Hemichromis bimaculatus:* the relieved parent swims away in a straight course, thus suppressing the following responses of the young (*after Baerends and Baerends, 1948*)

Baerends and Baerends [5] have shown that this prevents the young from swimming away with the parent who is off-duty.

It is interesting to see how human beings, on the level of reason, have evolved similar ceremonies. 'Greeting', whatever its psychological basis may be, often has the function of appeasement, of suppressing aggressiveness and related reactions, and thus opening the gate for further contact.

The function of isolation is probably of much less significance in the sphere of parent-offspring relationships than it is in mating. It is of course necessary to prevent the parents from extending their care to the young of other species, because this would be at the expense of their own offspring. Parental behaviour however is usually conditioned to the place where the young are. In species where the young wander with the parents, such as gallinaceous birds, isolatory functions seem to be more important; it is very probable that the distinctive head patterns of the chicks of this group serve this function. Voice too may in many species help to focus parental care on their own species.

The need for defence of the helpless brood introduces a new category of social behaviour. In many species the young are camouflaged. Camouflage however helps only if combined with stillness. Foraging and begging, on the contrary, require movement. Therefore, special behaviour has evolved in many species by which the parents can stimulate the young to 'crouch'. Thus the alarm-call of parent Blackbirds (*Turdus merula*) suppresses begging in the nestlings and makes them

PLATE 5

Young Herring Gulls in submissive posture

A 'crèche' of the King Penguin in South Georgia

crouch in the nest. How deep-rooted this reaction in the young is was made annoyingly clear to me when I tried to carry out some tests on the stimuli releasing begging in the young.[107] As soon as the parents called the alarm as a reaction to our presence near the nest the young birds became indifferent to our most tempting morsels. The reaction is still more fully developed in species whose young move about; usually such chicks have cryptic colours. Very young Herring Gulls simply crouch in the nest when the alarm sounds. When they grow up, they learn to know special hiding-places in the cover round the nest, and each runs to its shelter and crouches there when the parents call the alarm. Other species rely less on camouflage and have specialized on defence by the parents. The young of such species seek shelter near the parent, and at the side away from the predator, in the 'danger shadow'. This is found in many ducks and geese, and also in some fowl. Cichlids have developed the same type of response, of course quite independently.

The case of the Herring Gull is a good illustration of the need for distinguishing between the timing and the orienting function of alarm-calls. The alarm-call of the parents times the response; it releases in the young the desire to hide. However, the parent cannot tell them where the predator is, nor where to hide. This was made clear to me when I was taking photographs from a hide in a Herring Gull colony. The hide had been standing at the same spot for so many days that it had been accepted by all gulls, old and young, as part of the scenery. The parents used its roof as a look-out post, the chicks hid in it in case of danger. One day I made a careless movement while sitting in the hide, which was seen by one of the parents. It promptly called the alarm, and walked away from the hide. The young, aroused by the alarm-call, ran for shelter. My hide being their particular shelter, they all entered the lion's den and crouched at my feet.

Apart from relations between parents and young there are relations between male and female which enable them to divide their duties. For instance, in the Lapwing (*Vanellus vanellus*) the female does the incubating while the male stands on guard. His task is to attack predators, and to warn the female when a predator is approaching. Upon his warning call the female runs away from the eggs, entrusting their protection to their cryptic coloration. After she has run for some fifty yards, she flies up, and often joins the male in attacking the predator. In other species both male and female incubate. Here again there is a problem of timing. In many species the eggs are never left alone. How is the sitting partner prevented from abandoning the eggs before the mate arrives? This is done through a relief ceremony, for which the sitting partner has to wait, and without which it is very difficult for it to leave. When the non-brooding partner of a Herring Gull pair has been foraging for some hours its brooding urge begins to grow, and it flies to the territory. Here it collects some nest material, and walks up to the nest. Often it utters the 'mew-call', the same call it uses before feeding the young. It is this approach, with this call and this behaviour, which stimulates the sitting partner to get up. However, if its incubation drive is still too strong it may not react to even the most intense relief ceremony by the partner, and just remains on the eggs. If the partner fails to entice its mate from the eggs, it may try to remove it by force, and a quiet but determined struggle may be the result (Plate 4, lower fig.). Some species even know how to deal with the opposite situation. When a pair of Ringed Plovers is disturbed at the nest, it sometimes happens that even after the predator has gone away, the birds are still unwilling to go back to the nest. In such cases, the male can be seen to drive the female to the nest.[49]

A still more complicated situation arises when two successive broods are 'telescoped', that is to say when a new brood is begun before the young of the previous brood are independent. This happens regularly in the Nightjar, and often in the Ringed Plover. In the Nightjar the parents divide their duties strictly; the male stays with the chicks and the female sits on the new eggs.[44] In the Ringed Plover, both parents take turns both at sitting and at leading the young (Fig. 28), and every few hours they change over.[49] This again is timed by a special type of behaviour, by which one bird signals the other to change over.

The alarm-calls by which the parents warn each other are the same as those which influence the young. The partner's reactions to it, are, however, different. In some species the female crouches on the nest. This is the rule in species breeding in open country where the female is cryptically coloured, such as many Ducks, Nightjars, Curlews, and Pheasants. Other species behave like the Herring Gull and leave the eggs or young, in the meantime attacking the predator. The attack

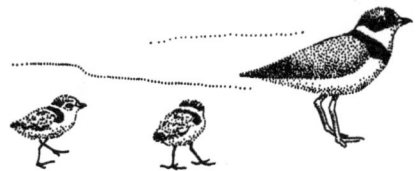

FIG. 28.—Ringed Plover with chicks

may be the affair of the individual pair, or it may be a truly social attack. The members of a Jackdaw colony for instance often attack in force. Each bird is not only reacting to the alarm-call of all others, but joins the others in the attack even though the predator may not be particularly near its own brood. In Common Terns I observed individual attacks on human intruders but communal attacks on a Stoat.

In many birds it is essential that both partners know the site of the nest, and there are numerous types of behaviour which point out the nest site to the partner. When male and female select the nest site together, there are ceremonies in which the birds join at the nest site. Thus Herring Gulls sit down on the selected spot, and take turns sitting in the nest and scraping out a nest pit with the legs. In many hole-breeding birds, the male, arriving on the territory before the female, selects the nest hole, and calls the female's attention to it by a special display. The male Redstart, for instance, has various ways of making itself conspicuous at the nest entrance (Fig. 29), making full use of its gaudily coloured head and of its red tail.[11] The male Kestrel sails down to the nest in a ceremonial way within view of the female.[90]

A special case was observed in the Red-necked Phalarope (*Phalaropus lobatus*). Here the female is the more conspicuously

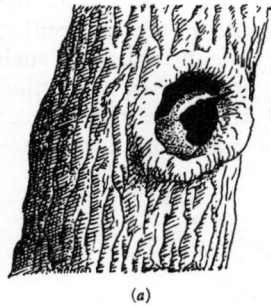

(a)   (b)

FIG. 29.—Two ways in which a male Redstart advertises its nest-hole to a
female. (a) Showing its gaudily coloured head. (b) Displaying its red
tail. (*Modified from Buxton, 1950*)

coloured partner, and the male has a dull, cryptic plumage.
The female selects and defends a territory, sings, and attracts
a male. The male alone incubates and guards the young. Since
the female is feminine enough to lay the eggs, she must be able
to point out the nest to the male. This is done by a revival of
her song just before she is going to lay an egg. The male cannot
resist the song any more than he could when he was initially
attracted by her at his arrival, and he follows her immediately.[93]
She then goes to the nest and lays the egg in the presence of the
male. It is this ceremony which guides the male to the site
where the female deposits the objects of his future care.

GROUP BEHAVIOUR

Many animals aggregate in groups larger than families. Such
groups may be composed of several families, such as a flock of
geese or swans, or they consist of individuals which are no
longer united by family ties. The benefits which individuals
derive from grouping together can be of various kinds. Among
these, the defence against predators is the most obvious one.
Members of a flock of higher animals warn each other in case
of danger, and the flock as a whole therefore is as alert as the
most observant of its individuals. Moreover, many animals join
each other in communal attack. These activities are mainly
found in the higher animals, but lower down in the scale we

find numerous other functions of flocking. Allee and his co-
workers [1, 2, 114] have experimentally demonstrated a great
number of these social benefits. Goldfish, for instance, eat more
when in a group than when living in isolation. They also grow
more rapidly; this is not dependent on increased food-intake
alone, but on other factors as well: it happens also when each
isolated goldfish gets the same amount of food as each of those
living in the group. The marine flatworm *Procerodes* withstands
fluctuations in salinity better when living in groups than when
isolated. The achievements of Cockroaches in orientation tests
are better when they are living in groups of two or three than
when they are kept in isolation. The advantage of crowding in
Daphnia was shown by Welty to lie in a reduced vulnerability
to predation. This is due to a 'confusion effect' on the predator;
when Goldfish were confronted with a very dense flock of
Daphnia, they were continuously tempted to turn away from
one Daphnia to another before they had snapped up the first.
Their total intake therefore was less than when feeding on a
group of moderate density. Crowding in *Vanessa io* Caterpillars
protects them from songbirds such as Redstarts; the latter were
consistently observed to leave groups of caterpillars alone but
pecked up every one that crawled away from the group.[64]

It is thus obvious that group life offers advantages of many
kinds to the individuals and thus to the species. Here again we
may ask: how does behaviour contribute to this beneficial
result?

First, the individuals must come together and stay together.
This can be effected by signals, acting upon various sense
organs of the reactors. In birds, these signals are usually visual,
or auditory, or both. The wing specula of ducks and geese,
which are brightly coloured and differ from one species to
another, have been shown to serve this function. Heinroth
found in the Berlin Zoo, where Anatidae of many parts of the
world are kept together, that ducks and geese (which often
react to a flying bird by flying up and joining it) do so most
readily when the flying bird has a speculum resembling their
own, irrespective of systematic affinity.[30] The conspicuous
and specific rump patterns of so many birds, notably waders,

undoubtedly serve the same function. The call-notes of many songbirds, such as Fringillidae and tits, serve to keep the group together; each individual is attracted by the songs of its own species, a fact which can easily be established by watching the behaviour of a bird that has strayed from the flock.

Many fish react mainly visually to each other, but smell plays a part in some species. Minnows for instance react to the scent of their own species.[113] They can even be trained to distinguish between the odours given off by different individuals,[27] but whether this type of individual recognition plays a part under natural conditions is unknown.

The social behaviour of higher animals extends beyond mere aggregating. In several species they co-operate more closely. As described in Chapter I, Sticklebacks react to the sight of another Stickleback eating by a tendency to start eating themselves. This effect is known as 'sympathetic induction' or 'social facilitation'; it has been observed in many species and occurs not only in the sphere of eating but with other instincts as well. When one bird of a flock gives signs of being alarmed, the others become alarmed as well. Sleep is another 'infectious' behaviour pattern. Even walking and flying are synchronized in this way; when some members of the flock show the intention movements of walking away, the others may join. A sudden take-off immediately makes the whole flock follow. The advantage of all these types of social facilitation is obvious; it synchronizes the activities of the members, and thus prevents them from scattering in the pursuit of diverse functions.

Most of these relationships depend on the tendency in each individual to react to the movements of the others. This tendency is highly developed; social animals are sensitive to even the slightest signs, to movements of very low intensity. These low-intensity movements, such as half-hearted, incipient walking or jumping, are called intention movements. Many social signals are clearly derived from such intention movements. Some social signals are very specialized. When a Jackdaw takes wing, it watches the other members of the group closely. If they do not take off, it either returns to them and gives up its attempts for the time being, or it entices them to join it.[54] This

FIG. 30.—Wagtails mobbing a Sparrow Hawk

it does by flying back to the individual(s) still on the ground, and gliding low over it, quickly shaking its tail while doing so.

Another type of social co-operation is the communal attack. This again is best known in birds. Many species such as Jackdaws, Terns, and various songbirds 'mob' a predator. They may gather in the bushes round a sitting Sparrow Hawk, or Little Owl, or above a prowling cat. This behaviour can often be watched in the House Sparrow. Or they may fly in a dense cluster above a flying Sparrow Hawk, trying to keep well above it, and now and then swooping down on it (Fig. 30).

Such behaviour may be released in all individuals at the same time because they see the predator simultaneously. If it is spotted by only one of them, however, this one calls the alarm and thereby warns the others. These alarm-calls are a clear example of an activity which serves the group but endangers the individual.

Such social attacks have various functions. If the predator is only moderately hungry, it can often be seen to hurry away as soon as the attack develops. When a Sparrow Hawk is really hungry, that is when it is hunting intensively, mobbing does not disturb it much. It nevertheless distracts part of its attention from detecting other prey. Even clustering together without mobbing, such as is done by Starlings or waders pursued by a Peregrine Falcon, has survival value; a swooping Peregrine takes care to select birds that are isolated from the flock;

because of its tremendous speed it could easily damage itself by swooping right through a dense flock.

The alarm-signal need not always be visual or auditory; it is known to be of a chemical nature in many social fish. When a Pike or a Perch snatches a Minnow from a school, the other Minnows scatter, and do not return to the vicinity. They remain on the alert for a long time, and dash into cover at the slightest sign of a predator. This is due to an olfactory response to a substance which was released from the skin of the killed Minnow. The reaction can be released in tame Minnows in an aquarium by mixing an extract of freshly cut Minnow skin with the food. The substance is specific, and so are the reactions to it, each species reacting to its own 'fright substance' only.[25]

CHAPTER IV

FIGHTING

WHEN an animal is cornered by a predator, it will often fight. This type of fighting, the defence against a predator, will not however concern us here, because it usually does not involve animals of the same species. Nor is it as common as the fighting of animals which is directed at individuals of their own species. Most of this intraspecific fighting is done in the breeding season, and is therefore called reproductive fighting. Some fighting has to do with dominance relationships in the group and is not linked with the breeding season.

REPRODUCTIVE FIGHTING

Different species fight in different ways.[63] Firstly, the weapons used are different. Dogs bite each other, and so do gulls, and various fish. To that end alone the male Salmon develops a formidable jaw. Horses and many other hoofed animals try to kick each other with the forelegs. Deer measure

FIG. 31.—Fighting Red Deer

their relative strength by pushing against each other with their antlers. Waterhens can be seen to fight all through the spring in many parks. They throw themselves halfway on their backs, and fight with their long-toed feet. Many fish fight by sending a strong water-jet towards the opponent by means of vigorous sideways tail-beats. Although they do not actually touch each

other, the movement in the water caused by the tail-beats gives a powerful stimulus to the opponent's highly sensitive lateral-line organs (Fig. 32). Male Bitterlings develop horny warts on the head in spring, and try to butt each other with the head.

FIG. 32.—Tail-fighting in fish (*after Tinbergen, 1951*)

Secondly, although so much fighting goes on all through the spring, it is relatively rare to see two animals actually engaged 'in mortal combat' and wounding each other.[103] Most fights take the form of 'bluff' or threat. The effect of threat is much the same as that of actual fighting: it tends to space individuals out because they mutually repel each other. Some instances of threat display were given in Chapter I. Its variety is almost endless. Great Tits threaten by facing each other, stretching the head upward, and swinging slowly from side to side, thus displaying the black-and-white head pattern.[95] Robins threaten by displaying the red breast, turning it to the opponent and then turning slowly right and left alternately (Fig. 33). Some Cichlids display the gill covers by raising them while facing the enemy. In *Cychlasoma meeki* and in *Hemichromis bimaculatus* these gill covers are adorned with very marked black spots, bordered by a golden ring; the threat display shows them off beautifully (Fig. 34).

Not all threat is visual. Many mammals deposit 'scent signals' at places where they meet or expect rivals.[29] Dogs urinate to that pur-

FIG. 33.—Threat display of the English Robin (*after Lack, 1943*)

pose; Hyaenas, Martens, Chamois, various Antelopes and many other species have special glands, the secreta of which are deposited on the ground, on bushes, tree stumps, rock, &c. (Fig. 35). The Brown Bear rubs its back against a tree, urinating while it does so.

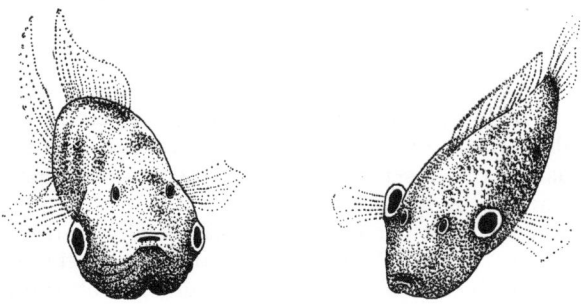

FIG. 34.—Frontal threat display in *Cychlasoma meeki* (*left*) and *Hemichromis bimaculatus* (*after Tinbergen, 1951*)

Sounds may also have a threat function. All the calls, mentioned in Chapter II under the collective heading 'song', do not merely attract females, but serve to repel males as well.

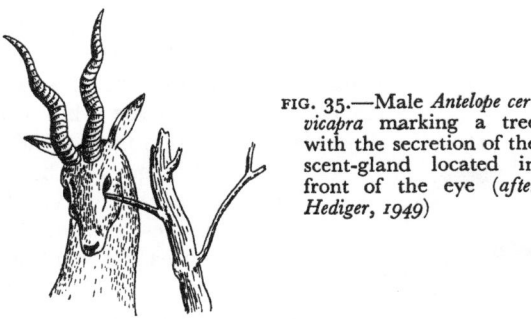

FIG. 35.—Male *Antelope cervicapra* marking a tree with the secretion of the scent-gland located in front of the eye (*after Hediger, 1949*)

THE FUNCTIONS OF REPRODUCTIVE FIGHTING

Reproductive fighting is always aimed at a special category of individuals. In most species it is the males which fight, and they attack exclusively, or mainly, other males of the same species. Sometimes male and female both fight; when that happens there is often a double fight: male attacking male, and female fighting female. In the Phalaropes and some other species of birds it is the females who fight, and they again attack mainly other females. This all shows that fighting is aimed at reproductive rivals.

Further, fighting, and threat as well, tend to prevent two rivals or competitors from settling at the same spot; mutual hostility makes them space out, and thus reserve part of the available space for themselves. An examination of what is essential in this reserved space will help us understand the significance of fighting.

The fighting of each individual is usually restricted to a limited area.[33, 94] This may be the area round the female(s) as it is in Deer and many other animals. Bitterling males defend the area around a Freshwater Mussel against other males (Fig. 36); to this Mussel they attract a female. They induce her to lay her eggs in its mantle cavity, where they will develop, leading a parasitic life. Carrion Beetles of the genus *Necrophorus* defend carrion against rivals. The defence in all these cases not only concerns the central object itself, but also a certain area round it; rivals are kept at a considerable distance. In the species mentioned it is easy to see what the central object is: when a doe moves and walks off the male will go with her; it always fights in her vicinity. When the Mussel moves, the male Bitterling shifts the area it defends along with it. In most species however the defended area does not move; the male settles down on a chosen spot, and defends a territory. This is known in many animals; territorial fighting and threat can be watched in every garden, for Robins, Chaffinches (Fig. 37), and Wrens, to mention only a few species, are renowned fighters. It is possible to understand the

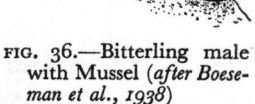

FIG. 36.—Bitterling male with Mussel (*after Boeseman et al., 1938*)

significance of such a territory when the fighting is centred on one particular part of it. Thus in many hole-breeding birds the fighting is particularly furious when intruders come near the hole. In most species, however, there is no such concentration on a particular part of the territory, and here the significance of territory is less easy to understand. It has been suggested that the territory of many songbirds might be useful as a

FIG. 37.—Fighting male Chaffinches

reservoir of food for the young. This would enable the parents to collect a certain basic quantity of food near the nest, which would mean that the foraging trips could be short. Since newly hatched songbirds have to be brooded in order to remain warm and ready to gape for food, the territory might be a help in keeping foraging trips, and thus intervals between bouts of brooding, as short as possible. The length of the intervals between bouts of brooding might, on unfavourable days, be critical. Opinions about the value of this argument differ, however.

In ground-breeding birds such as Gulls, Terns, Lapwings, &c., spacing-out appears to be part of the defence of the brood against predators. There is evidence to show that too dense a concentration of prey such as eggs or chicks makes predators specialize on them; that is the main reason why camouflaged animals are as a rule solitary and well spaced out.[102] In birds like Gulls, where the brood is camouflaged, territorial fighting has the effect of keeping the individual broods reasonably far apart. Here again the conflict between two interests has resulted in compromise: social nesting has certain advantages (as we have seen in Chapter III); so has spacing-out. The various species of Gulls and Terns have each arrived at a compromise which give them some, though not complete, benefit of both tendencies.

Concluding, it is clear that reproductive fighting serves a function. It results in a spacing-out of individuals, thus

6

ensuring each of them the possession of some object, or a territory, which is indispensable for reproduction. It thus prevents individuals sharing such objects, which would in many cases be disastrous, or at least inefficient. Too many Bitterling eggs in a Mussel will result in a low ration for each. When many males would mate with one female instead of securing females of their own this would be a waste of germ cells. Two broods of Starlings in one hole may be fatal to both broods. Spacing out makes the individuals utilize the available opportunities.

THE CAUSES OF FIGHTING

Our next problem is: what makes the animals fight in such a way as to promote these functions? What makes them fight only when it is necessary, and only at the place where it is necessary? How does the animal select its potential rival amongst the multitude of other animals it meets? Since fighting endangers the individual (because it makes it vulnerable to attack by predators), and since it may endanger success in reproduction because unlimited fighting might leave the animal little time to do anything else, restriction of fighting to situations in which it can serve its functions is of vital importance. These problems are rather similar to those discussed in relation to mating. In order to confine fighting to the actual defence of territory, the Mussel, the female, &c., the animal will have to react specifically to these situations. Further, fighting must be timed, that is, confined to those moments when there is a rival to be driven off. Finally, it must not be wasted on other species, except when these are rivals. As we will see, many of the outside stimuli responsible for these various aspects of co-ordination are provided by the rival. Since, moreover, most of these stimuli serve more than one of these functions, I will not divide this treatment into sections according to the function served as rigidly as I did in the chapter on mating.

As we have seen, restriction to a certain locality is one of the most obvious characteristics of fighting. When a male Stickleback meets another male in spring, it will by no means always fight. Whether it does depends entirely on where it is. When in its own territory, it attacks all trespassing rivals. When outside its territory, it will flee from the very same male which it would

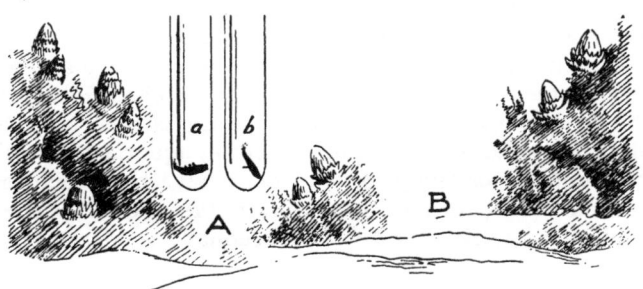

FIG. 38a.—A test on the dependence of attack on territory: Male *b*, owner of territory B, is brought in a glass tube into territory A of male *a*; the latter attacks while the former tries to escape

attack when 'at home'. This can be nicely demonstrated in an aquarium, provided it is large enough to hold two territories. Male A attacks male B when the latter comes into A's territory; B attacks A when A trespasses. Usually neither male

FIG. 38b.—The same males in territory B: male *b* attacks while male *a* flees

trespasses voluntarily on to strange territory, but one can easily provoke the situation by capturing the males and putting each of them in a wide glass tube. When both tubes are lowered into territory A, A will try to attack B through the double glass wall, and B will frantically try to escape. When both tubes are moved into territory B, the situation is completely reversed (Fig. 38).

How the territory stimulates the male to fight has rarely been studied in any detail. It can of course only be found out by experimentally moving the territory or parts of it and seeing whether the male adjusts its fighting to the changed situation. This is of course difficult in birds because of their large territories, but small fish that can be kept in aquaria offer unique opportunities for study. Several cases have been reported of birds extending their territory after the female had started to build a nest outside the territory originally staked out by the male.

It seems certain that territories are selected mainly on the basis of properties to which the animals react innately. This makes all animals of the same species, or at least of the same population, select the same general type of habitat. However, the personal binding of a male to its own territory—a particular representative of the species' breeding habitat—is the result of a learning process. A male Stickleback is born with a general tendency to select a habitat in shallow water with liberal vegetation, but it is not born with the tendency to react to a particular plant here and a pebble there. It shifts its territory when we move these landmarks because it has been conditioned to them. This can be seen from the fact that when it breeds two or three times in succession, it often moves to new territories. In each of them it orients itself to landmarks.

Species which react to a special object, such as a hole, or, as the Bitterling does, to a Mussel, probably react innately to it, and as a consequence react to only few 'sign stimuli' provided by it. The Bitterling,[8] for instance, reacts only to a minor extent to the visual stimuli provided by the Mussel; the main stimulus is the exhalation current sent out by the Mussel; the fish reacts both to the movement of the water and to its chemical properties (Fig. 39).

Stimuli from the territory to which the animal reacts either innately or as an added result of conditioning, makes the animal confine its fighting to the territory.

The gross timing of the attack is again a matter of outside factors. As in mating, the first, very crude timing depends on sex hormones. Fighting appears as a consequence of gonadal growth which, in its turn, through the pituitary gland, depends on rhythmic factors such as day lengthening in the case of many animals of the northern temperate zone. The more accurate timing however is again a matter of reaction to signals. Signals from the rival release fighting when the latter comes too near the territory, or whatever the defended object may be. These signals always have a curious double function. When displayed by a stranger, they draw the attacker to it. When displayed by an attacker on its own ground, they intimidate the stranger. When experimenting with models one can release both responses, dependent on the place—inside or outside the territory—where they are presented. In both cases they serve to space out the species, and, since the responses are specifically released by these displays, not

FIG. 39.—Bitterlings react most intensively to an empty Mussel shell when water is led through it in which live Mussels have been kept (*after Boeseman et al., 1938*)

by threat displays of other species, they tend to confine hostilities within one species.

These stimuli have been analysed in various species by experiments with models. The male Three-spined Stickleback, while showing some hostility towards any trespassing fish, concentrates on males of its own species. Models of males release the same response, provided they are red underneath. A bright blue eye and a light bluish back add a little to the model's effectiveness, but shape and size do not matter within very wide limits. A cigar-shaped model with just an eye and a red underside releases much more intensive attack than a perfectly shaped model or even a freshly killed Stickleback

which is not red (Fig. 40). Size has so little influence that all
males which I observed even 'attacked' the red mail vans

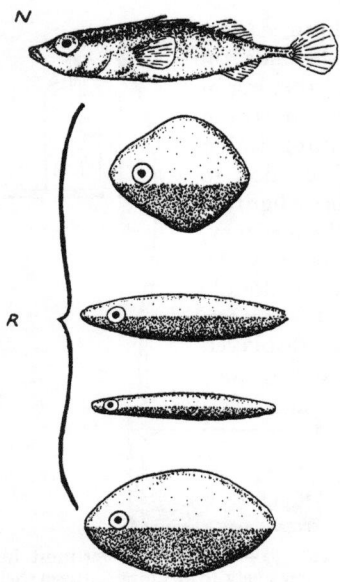

passing about a hundred
yards away; that is to say
they raised their dorsal spines
and made frantic attempts to
reach them, which of course
was prevented by the glass
wall of the aquarium. When a
van passed the laboratory,
where a row of twenty aquaria
were situated along the large
windows, all males dashed
towards the window side of
their tanks and followed the
van from one corner of their
tank to the other. Because
models of three times stickleback size, although releasing a
similar attack as long as they
were not too close, were not
actually attacked when
brought into the territory, it
seems that the angle sub-
tended by the object is important and this must be the
reason why the distant mail
vans were attacked.

FIG. 40.—Models used in experiments on the release of fighting
in male Three-spined Stickleback.
A perfectly shaped silvery model
(N) is rarely attacked, while crude
models with red undersides (R)
are strongly attacked (*after Tinbergen, 1951*)

Apart from colour, be-
haviour may release the
attack. A male Stickleback viewing a neighbour from afar
often adopts a threat posture, a curious vertical attitude, head
downward (Fig. 41). The side, or even the underside, is turned
towards the opponent, and one or both ventral spines are
erected. This posture has an infuriating effect on other males,
and we can intensify a male's attack on a dummy by presenting it in this posture.

Similar observations have been made on the Robin. When

a male Robin has staked out his territory, the sight of another Robin in this territory releases attack or threat. Lack showed that the red breast is the releasing factor more than anything else.[47] When he put up a mounted Robin in an occupied territory, this was postured at by the owner. Even a small cluster of red feathers was sufficient to evoke posturing (Fig. 42). And, just as in the Stickleback a very crude red model was more effective than a perfectly shaped but silvery model, so to the Robin these few red feathers had more meaning than an entire mounted immature Robin, which had all the form-features of its species but had a brown instead of a red breast. It is remarkable how similar are the functions of the red breast of the male Stickleback and those of the Robin's red breast. We will see that comparable signal-

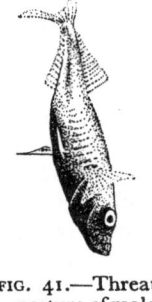

FIG. 41.—Threat posture of male Three - spined Stickleback (*after Tinbergen, 1951*)

ling systems have convergently developed in animals of other groups as well.

In the Robin, the signalling is not entirely visual however. Robins hear each other over far greater distances than they see each other. In particular, the song of a Robin arouses the owner of a territory and sets him off to find the singer. The real attack therefore occurs in at least two steps: a male first flies in the direction from which it hears another male singing; it then looks round and is roused by the red breast of the intruder to posture or attack.

FIG. 42.—An experiment on fight-releasing stimuli in the Robin: a mounted immature Robin, which has brown breast feathers (*left*) is rarely attacked, while an isolated tuft of red feathers (*right*) is postured at (*after Lack, 1943*)

In many other species song has this same function; it is a

'badge of masculinity' and releases fighting in territory holders. As I said earlier, such a badge of masculinity, when demonstrated by a male on its own territory, drives off intruders. This can easily be seen in the field without experimentation, because it often happens that a singing male cannot be seen, when for instance it is hidden by a tree or a shrub. It is fascinating to watch the intense reactions of other birds to such concealed singers. Trespassers are personifications of a bad conscience; territory owners those of righteous indignation.

In the Herring Gull, though the sexes have the same coloration, aggressiveness is mainly found in the males, and directed against other Herring Gull males. Male Herring Gulls however do not sing, nor does any of their calls particularly arouse

FIG. 43.—Female (*left*) and male American Flicker (*after Noble, 1936*)

other males. Nor do the males have conspicuously coloured parts that release fighting in other males. Their behaviour however does. Threat postures and nest building movements particularly draw the attention of other males and elicit a hostile response from them.

Other species again resemble the Stickleback in that the male's bright colours act as their badge. This for instance was found to be the case in the American Flicker (*Colaptes auratus*), a Woodpecker in which the male has a black patch at the corner of the mouth (the so-called moustache) which the female has not (Fig. 43). When the female of a pair was captured and given an artificial black moustache she was attacked by her own mate. After recapture and removal of the moustache she was again accepted.[67]

Male Shell Parrakeets (*Melopsittacus undulatus*) differ from females in the colour of the cere, which is blue in males, brown in females (Fig. 44). Females with the cere painted blue were attacked by males.[12]

A most remarkable parallel was found in a group as different as the Cephalopods. Males of the Common Cuttlefish, *Sepia officinalis*, have a brilliant visual display at the mating time. On meeting another Sepia they show the broad side of their arms, and by co-ordinated action of their chromatophores develop a conspicuous pattern of very dark purple and white (Fig. 45). The fighting of the males is a response to this male

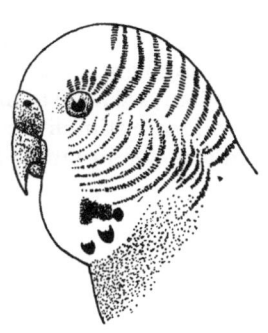

FIG. 44.—Head of Shell Parrakeet (*after Tinbergen, 1951*)

display; experiments with plaster models showed that the display acts visually; both shape and colour pattern contributing to the release of attack.[91]

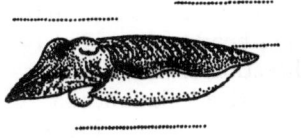

FIG. 45.—Male Cuttlefish at rest (*below*) and displaying (*above*) (*after L. Tinbergen, 1939*)

Lizards behave much as Cuttlefish do.[38, 42, 66, 68] The males have special movements which serve to display specific male colours. The American Fence Lizard (*Sceloporus undulatus*) has procryptic colours on the back. The underside of the males, however, is a clear blue. This colour is not visible until the male displays, as it does in spring upon meeting another Fence Lizard. It then takes up a position in front of the other and at right angles to it, and compresses its body laterally so that the blue underside becomes visible from the side (Fig. 46). By changing the colour of males and females with lacquers, Noble showed that the blue belly releases fighting in territory holding males.[66, 68]

So far, I have been reviewing examples of stimuli responsible mainly for the timing of fighting. In most of these cases they direct the fighting at the same time. However, as in mating behaviour, we have to distinguish between these two functions, for there are stimuli contributing to one and not to the other. In ducks, for instance, females make special movements and

FIG. 46.—Male Fence Lizard in display (*after Noble, 1934*)

calls urging their mates to attack other males. The calls merely raise the male's aggressiveness, but by special head movements the female points out to her mate the male to be attacked.[56] This can easily be seen in the tame and half-tame Mallards living in our parks: the female swims from an 'accosting' male to her own mate, repeatedly pointing her head with a sideways movement over her shoulder in the direction of the stranger.

The third problem, related to reproductive isolation, that of confining the fighting to members of the same species, has also been covered by the examples given. Again, as with the signals which play a part in mating behaviour, the fight-evoking signals are specific, and very different even in closely related species if they are living in something like the same habitat. Yet one gets the impression that interspecific fighting is not as rigorously eliminated in evolution as interspecific mating. It seems, so far as the evidence goes, that what little interspecific fighting occurs is directed at species superficially resembling the species of the attacking animal. 'Erroneous' attacks occur because a strange species happens to present some of the sign stimuli normally releasing attack. In some cases, however, fighting is clearly aimed at other species because they are competitors for the same 'indispensable object'. Thus Starlings and Tree Sparrows are known to drive other species from nesting holes.

THE PECK-ORDER

Animal species living in groups sometimes fight over other issues than females or territories. Individuals may clash over food, over a favourite perch, or possibly for other reasons. In such cases, learning often reduces the amount of fighting. Each individual learns, by pleasant or bitter experience, which of its companions are stronger and must be avoided, and which are weaker and can be intimidated. In this way the 'peck-order' originates, in which each individual in the group knows its own place. One individual is the tyrant; it dominates all the others. One is subordinate to nobody but the tyrant. Number three dominates all except the two above it, and so on. This has been found in various birds, mammals, and fish. It can easily be seen in a hen-pen.

The peck-order is another means of reducing the amount of actual fighting. Individuals that do not learn quickly to avoid their 'superiors' are at a disadvantage both because they receive more beatings and because they are an easier prey to predators during fights.

The behaviour leading to peck-orders has some interesting aspects. Lorenz found in Jackdaws that when a female of low 'rank' got engaged to a male high up in the scale, this female immediately rose to the same rank as the male, that is to say that all the individuals inferior to the male avoided her though several of them had been of higher rank than she before.

The American literature contains many valuable contributions on the problems of peck-order.[1, 2] In many of these papers, however, peck-order is claimed to be the only principle of social organization. This leads to distorted views; peck-order relationships form only one category among the numerous types of social relationships in existence.

CHAPTER V

ANALYSIS OF SOCIAL CO-OPERATION

RECAPITULATION

IN the preceding chapters I have tried to show that social co-operation serves a great variety of ends. Mating behaviour is not merely the act of coition, but is preceded by long preliminaries. These preliminaries, or courtship, have very distinct functions. It is necessary that the two partners are brought together. Their activities must be synchronized. The reluctance against bodily contact must be overcome. Interspecific matings must be prevented. The female must appease the male's aggressiveness. We have seen that all these functions are served by a signalling system, by which one individual can influence the other's behaviour. In family life, the behaviour of the parents has to be co-ordinated so that they take turns in guarding the eggs or young. When the young are to be fed, or when they must be warned against a predator, close co-operation, often involving mutual signalling, is necessary. Several of the relationships of family life extend beyond that into group life, and here again we found that co-operation was based on signalling. Finally, I argued that fighting, and especially reproductive fighting, although in some respects a disadvantage to the individual, is of great use to the species, because it effects spacing out and thus tends to prevent harmful overcrowding. Since actual fighting inflicts damage as well as effecting spacing-out, a signalling system such as exists in most species, where damage is reduced to a minimum while the intimidating effect is retained, is to the species' advantage. Threat display reduces fighting in two ways: if shown by an owner (of a territory, a female, a hole, etc.) it intimidates rivals. If shown by a trespasser, it marks the latter out for attack, and thus enables an owner to leave harmless intruders alone. Again, these functions depend on signals.

The signalling system has been studied in a number of cases.[99] Although much more work has still to be done, some general conclusions are already possible.

We have seen that the parent Herring Gull feeds the chicks by regurgitating some food, and presenting part of it to the young, keeping it between the tips of the bill. The young gull is first roused by the 'mew call' of the parent, then it pecks at the bill tip, clearly guided by visual stimuli, until it gets the food in its bill, when it swallows it. The various signals, auditory and visual, are given by the parent and reacted to by the chick. In discussing such signalling systems, I will call the individual presenting the stimuli the actor, and the individual responding to the stimuli the reactor.

THE ACTOR'S BEHAVIOUR

Our central problem is: what urges the actor to signal? What makes the parent gull call the young, and present food to it? Judging from our own behaviour, we would be inclined to think that the actor has a special end in view, and that it acts in order to attain that end. There is strong evidence indicating that such an amount of 'foresight'—which in some unexplained way controls our own behaviour to such a great extent—does not control the activities of animals. If there were such foresight, and such insight into the ends served by behaviour, we could not explain the numerous cases where animal behaviour does not reach its goal and yet the animal does nothing to remedy it. For instance, if alarm-calls were given with intent to warn other individuals, it would be incomprehensible why birds call the alarm with equal vigour whether or no there is another bird to be warned. Or, if parents were guided by insight into the function of brooding and feeding their young, songbirds parasitized by a Cuckoo would not let their own young die before their eyes after the young Cuckoo has thrown them out of the nest. Such behaviour, and there are numerous similar examples, can be shown to be due to relatively rigid and immediate responses to internal and external stimuli. A parent songbird cannot feed young when these do not beg for food.   It cannot brood them unless they are in the nest. On the other hand, a parent bird must give the alarm-call when it perceives a predator, irrespective of the presence of another bird to be warned.

Returning to the Herring Gulls, all the evidence leads us to conclude that the parent is reacting rigidly to an internal urge and to the stimuli from the nest site and from the young themselves. The rigidity of such behaviour shows plainly in the reactions of a parent gull to a dead chick. I have seen, more than once, a chick killed by a neighbouring gull. Although the chick's father and mother will furiously defend it as long as it is alive, they will devour it as soon as it is dead. They no longer hear the chick's calls, they do not see its movements, and that is sufficient to make it lose all significance as a chick and to become food.

There can be no doubt that this conclusion can be generalized. Except perhaps in the highest mammals, all signalling behaviour is immediate reaction to internal and external stimuli. In this respect there is a great difference between animals and Man. The signalling behaviour of animals can be compared with the crying of the human baby, or with involuntary expressions of anger or fear in humans of all ages. We know that such 'emotional language' in Man is different from deliberate speech. The 'language' of animals is of the level of our 'emotional language'.

Further, signalling behaviour is innate in probably all the cases that have been discussed here. This has been proved in a number of animals by raising them in isolation from other members of their species, so that they have no opportunity to see and imitate their behaviour. As a matter of fact, real imitation is now known to be extremely rare in animals. Yet it is always a surprise to see such isolated animals perform even complicated behaviour patterns such as building a nest, fighting an opponent, or courting a female for the first time in their lives. For instance, when I raised a Three-spined Stickleback in isolation from the egg stage on, it showed the complete fighting behaviour, and the complete chain of courtship activities when, after having reached sexual maturity, I confronted it with a male and a female. In this respect as well, animal 'language' differs from human speech.

In some cases we know something about the causes responsible for the particular type of the actor's behaviour. It has always struck observers that all kinds of 'display', whether courtship, or threat, or other types of signalling, consist of such grotesque behaviour. One general rule was established long ago: whenever conspicuously coloured parts of the body are used in display, they are always made clearly visible. Crests are raised, wings or tails lifted, bills opened widely whenever these parts of the body are gaudily coloured. The broad side of such bright parts is always turned towards the reactor. Many birds display beautifully coloured fans towards the female; collars, wings, or tail are displayed frontally or laterally. Fish spread their gill covers when threatening frontally, they raise all their fins when displaying laterally. Movement and structure co-operate to attain a maximum visual effect.

In several cases it is now known not only that display is a reaction to external and internal conditions, but also why the display takes the form it does. This is best known in threat and in courtship.

Analysis of the circumstances leading to threat has shown that it arises when two drives are activated simultaneously in the actor: the drive to attack and the drive to escape. In territorial conflicts it is easy to understand how this can happen: since a stranger intruding in the territory releases attack, and when outside the territory elicits escape, a territory-holder meeting a stranger just at the boundary of his territory is simultaneously roused to attack and to flee. This creates 'tension', or strong activation of two antagonistic drives, and in such circumstances so-called 'displacement activities' appear through which the thwarted drives find an outlet.[97, 104] The threat posture of a Three-spined Stickleback is such a displacement activity. When two males are engaged in a very intensive fight, their curious head-down threat posture develops into complete sand-digging, the first phase of nest-building. The thwarted attack and escape drives, whose motor patterns are antagonistic and cannot occur together, find an outlet through this movement. Other species behave similarly during boundary conflicts; but the displacement movements used are different from one species to another (Fig. 47). Thus Starlings and Cranes preen their plumage; tits show feeding movements, many waders even assume the sleep posture!

Returning to the Sticklebacks, the threat posture is not merely displacement sand-digging. Usually they turn their broad side to the opponent, and erect one or two ventral

FIG. 47.—Various displacement activities functioning as threat

Upper left: 'grass pulling' (displacement nest building) in fighting Herring Gull (*after Tinbergen, 1951*)

Upper right: displacement sleep in Avocet (*after Makkink, 1936*)

Lower left: displacement sleep in fighting Oystercatcher (*after Tinbergen, 1951*)

Lower right: displacement eating in fighting Domestic Cocks (*after Tinbergen, 1951*)

spines. This is part of the behaviour pattern of the activated drives themselves, they are elements of defence against an enemy. Any Stickleback does the same when cornered either by another Stickleback or by a predator such as a Pike. The attack drive also finds expression in the threat behaviour: threatening males bite furiously into the sand, much more so than when really digging sand while preparing a nest site. This biting reminds one of an actual attack on the opponent;

they do with the sand (the object of the sand-digging) as they would do with the opponent 'if they only dared'.

A similar threat movement is found in the Herring Gull. In Chapter I I described how fighting Herring Gulls tear grass or moss out of the ground. This is displacement-collecting of nest-material, which acts as a threat. It is different from genuine collecting of nest-material; a threatening gull pecks much more energetically in the ground than it would do if it were only picking up straws for the nest. Also, it selects roots, firmly attached bunches of grass and the like, and pulls at them for all it is worth. This again is exactly what it does to another Herring Gull when it really gets to grips with it.

Such displacement activities appear only when the tension is very high. In milder forms of conflict the threat behaviour usually takes the form of a combination of parts of the behaviour patterns of both underlying drives. Sticklebacks dash back and forth, alternately attacking mildly, and withdrawing again. A Herring Gull combines elements of both drives into one posture: the stretching of the neck, the downward-pointing of the bill, and the raising of the wings are part of fighting; they are preparations for the delivering of pecks and wing blows. When the opponent is near, the neck is more withdrawn, as an indication of a tendency to retreat. This 'upright threat posture' therefore is incipient attack, toned down by incipient retreat.

Courtship movements occur also in conditions of tension. But the underlying drives are different. The sex drive is always involved. It may, however, be thwarted by various conditions. We have seen that co-operation in sexual behaviour is often dependent on mutual signalling. Whenever an animal is waiting for a signal from the partner, and this signal is for some reason not forthcoming, the next reaction in the chain, which depends on this signal for release, cannot be given. This situation leaves the animal with a strongly aroused, but thwarted, sex drive. A displacement activity is the result. The movement by which a male Stickleback shows the nest entrance to a female has been shown to be such a displacement activity; it is performed as long as the male is waiting for the

7

female to enter the nest. The quivering, which releases the female's spawning, is also a displacement activity, performed while the male is waiting for the female to lay the eggs, which alone can elicit his sperm-ejaculation. These two activities are both displacement-fanning. Genuine fanning is the movement by which the male sends a water current into the nest and thus ventilates the contents; it is part of the parental behaviour pattern.

The male Stickleback's zigzag dance is due to another situation. It is due to the fact that the female activates two different drives in the male: she stimulates him to attack, and at the same time she stimulates him to lead her to the nest, which is purely sexual behaviour. Each 'zig' can be shown to be an incipient leading movement, each 'zag' is incipient attack.[106] The zigzag dance therefore is a combination of two incomplete movements, and is due to the activation of the attack drive and the sex drive.

These few examples show that at least some display behaviour consists of movements which are derived from other patterns. They are either combinations of elements of the underlying drives, or they are displacement activities, derived from entirely different parts of the behaviour pattern. Although such analyses have only been made in few species, there is reason to believe that most signalling movements are in reality such derived movements. For reasons which will be discussed in Chapter VIII, such derived movements cannot always be recognized at a glance; detailed comparative study is often necessary.

THE BEHAVIOUR OF THE REACTOR

Turning now to the reactor's behaviour, we find again that it is innate. The Herring Gull chick aims its pecking response at the parent's bill tip from the first, without having to learn it. The male Stickleback raised in isolation reacts to other males by fighting, to females by courting. It could not have learnt this. In other words, it is not only the capacity to perform these motor patterns that is innate, but their sensitivity to special releasing and directing stimuli as well.

PLATE 6

Herring Gull feeding chick

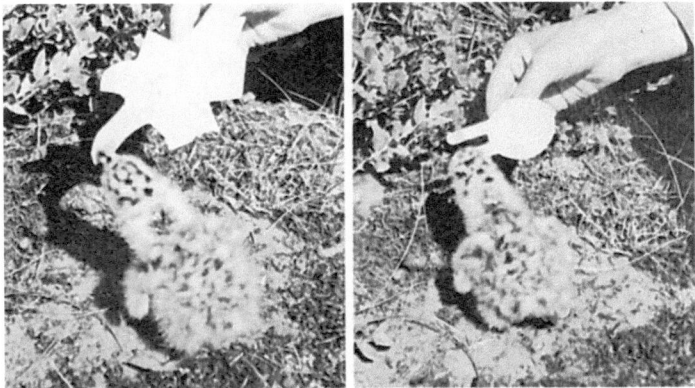

Two experiments with head models. Left: head of abnormal shape. Right: two 'bills' on one 'head'; chick pecking at lowest of the two bills

The responsiveness to signals has been the object of special study in a number of cases. Some of the results have already been described in the preceding chapters. We will now consider the begging response of the Herring Gull chick more closely, because here we know exactly to what stimuli the chick responds.[111] It is possible to release the begging response of a newly born, inexperienced chick by presenting it with a flat cardboard model of the parent's head. The chick responds to this just as well as to the real head (Plate 6). The bill tip of the adult Herring Gull bears a red colour patch which stands out quite conspicuously against the yellow background of the bill itself. When this red patch is absent in a model, the chick will respond much less vigorously than to the normal model with the red patch (Fig. 48). When these two models were presented in turn to a number of chicks, the average number of responses to the model without a red patch was only one-fourth of that to the normal model. Models in which there was a patch, but of colour other than red, released intermediate numbers of responses. This depended on the degree of contrast between the patch and the bill colour. In

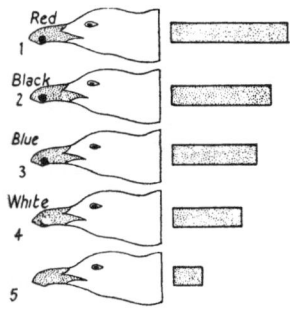

FIG. 48.—Models of Herring Gull heads with bill patches of various colours (1–4) and without patch (5). Columns to the right indicate frequencies of begging responses released by the models (*after Tinbergen and Perdeck, 1950*)

the same way, viz. by comparing the chick's responses to various models, it was possible to study the influence of the yellow colour of the bill. Surprisingly enough, the colour of the bill in the models did not make the least difference to the chicks, except that a red bill released twice as many responses as any other colour (Fig. 49). A bill in the natural yellow colour did not release more responses than did a white, a black, a green, or a blue bill. Neither did the colour of the head make any difference: one would expect that a white head would release more responses than a black or a green head, but that was not so. Nor did the shape of the head matter; it did not even make much difference when there was no head at all, but merely a bill. Yet the chicks can see the head very well, for they peck occasionally at the base of the bill, and even at the parent's red eyelids. When the chicks are hungry, there is just one thing to them that matters: the parent's bill with the red tip. In addition, the bill must be thin and elongate, it must point down, it must be as near the chick as possible, and as low as possible. But these are all the stimuli; everything else is irrelevant to the chick. It is remarkable how well the parent's behaviour and colour fit in with this, how they fulfil, so to speak, all the chick's expectations. The parent walks up to the chick, presents its bill in an almost vertical position, pointing the tip down, and it has a red blotch at the tip of the bill. This close correspondence between the characters of the parents and the stimuli to which the chick responds is amazing when we recall that the chick cannot 'know' what the parent looks like, or how it behaves.

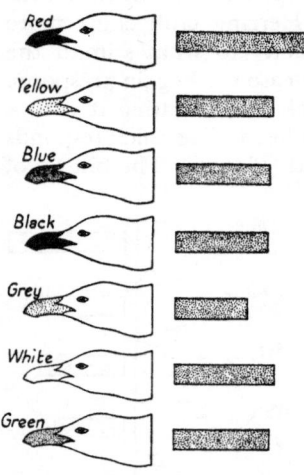

FIG. 49.—Models of Herring Gull heads with bills of various colours. Red releases more responses than any other colour, including yellow (*after Tinbergen and Perdeck, 1950*)

In many of the other animals that have been studied we find that the reactor, just like the gull chick, responds to a few selected stimuli provided by the actor. As we have seen, the fighting of the Robin is released by the red breast more than by any other bodily character. The male Stickleback's fighting is released by the red underside more than by anything else. The male 'moustache' of the Flicker overrides the influence of any other character, and so on. It seems as if such colours, shapes, calls, movements, have but one function: the release of fitting responses in the reactor. This idea was first clearly put forward by Lorenz,[55] who pointed out that social responses are often released by such features, which seemed to be specially adapted to this function. Such organs he called releasers. Lorenz described this concept of releaser in the following words: 'The means evolved for the sending out of key stimuli may lie in a bodily character, as in special colour design or structure, or in an instinctive action, such as posturing, 'dancing' and the like. In most cases they are to be found in both, that is, in some instinctive acts which display colour schemes or structures that were evolved exclusively for this end. All such devices for the issuing of releasing stimuli, I have termed releasers (*Auslöser*), regardless of whether the releasing factor be optical or acoustical, whether an act, a structure, or a colour.'

The evidence which is now accumulating through the work of a number of workers in this field seems to confirm Lorenz's hypothesis in the main. In very few cases is the evidence complete enough, and much more work remains to be done, but on the whole the principle of releasers seems to be a very useful one for understanding the mechanisms of social co-operation. The following review of releasers will be arranged according to the sensory modality involved, not according to the functions served.

REVIEW OF RELEASERS

Sounds play a part in those groups which have well-developed organs of hearing. We have already seen that males of many species attract females by a specific loud call, which, when it happens to strike human beings as beautiful, is honoured with the name of song. Experimental evidence of the influence of song as a releaser is scarce, and it would be worth while to use the many available records of birds' songs for experiments. It is probable that the 'bleating' of the male Snipe (Fig. 50), the rattling of the male Nightjar, the drumming of the male Woodpecker all have sharply defined functions, but it would be worth while finding out by experiment.

Another group in which calls and 'song' play a part is that of the frogs and toads. Of course we all know the croaking of male Common Frogs and of Common Toads. In subtropical and tropical regions many more noisy species are found and several of them have melodious voices so that one is less reluctant to apply the term 'song' to their performances than to the harsh noises of our frogs. Much remains to be learnt about the exact function of these amphibian songs, and of their other calls.

FIG. 50.—Male Snipe 'bleating'. The sound is produced by the vibrating outer tail feathers

Although there is some experimental evidence that the song of locusts and crickets has essentially the same functions as song in birds, we know next to nothing about the functions of all the other insect sounds. A courting male Grasshopper produces a series of different chirps; these sounds are typical of the species, and there is great regularity in the performance of each species. Sounds are further known in Cicadas, in ants, and in various other groups, but their functions are unknown.

Chemical signals, acting upon olfactory organs, are not uncommon either, but here again their function is

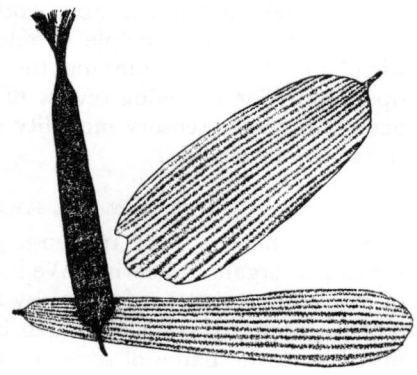

FIG. 51.—Two ordinary scales and one scent-scale of the male Grayling (*after Tinbergen et al., 1942*)

understood in only a few cases. I have mentioned already the attraction of male moths to females by scent, and the scent-advertising of territorial boundaries by mammals. Scent may also play a part in actual courtship as a persuasive, eliciting the female's co-operation. This is the function of the so-called scent scales in male Graylings. These scent scales are concentrated in a narrow strip on the upper side of the male's forewings. Their brush-like shape helps the secretions of the scent glands to evaporate into the air (Fig. 51). The climax of the male's courtship in front of the female is the 'bow': the male spreads the forewings and catches the female's antennae between them. The terminal olfactory organs, situated on the clubs of the female's antennae, are thus brought into contact with the scent field. Males whose scent scales have been brushed off and their

FIG. 52.—Mating Garden Snails. *Right:* a 'love dart' (*after Meisenheimer, 1921*)

bases covered with shellac are less successful in courtship than intact males, which, as a control, have shellac on other parts of the wing.[108]

Touch stimuli also play a part in social co-operation. When a male Stickleback has led the female to the nest, she enters it, and is then ready to lay eggs. This, however, requires a tactile stimulus from the male. His 'quivering' serves this purpose.

Another example of touch stimuli in courtship is provided by the mating of Garden Snails (*Helix pomatia*) (Fig. 52). These snails, being hermaphroditic, have an entirely mutual courtship. It consists of a series of postures and movements ending in coition. Szymanski[88] has been able to release the complete courtship behaviour in snails by imitating stimuli, normally provided by the partner, by gently touching a snail with a brush. This 'tactile courtship' culminates in a very vigorous stimulus: the 'love dart', a sharp calcareous arrow, is thrust into the partner's body, and this leads to coition.

As already mentioned, the threat display of many fish involves tactile stimuli of a special kind, acting upon the lateral line organs.

The courtship of various newts seems to be a series of signals of a visual, tactile, and chemical nature.[61, 110] The male of

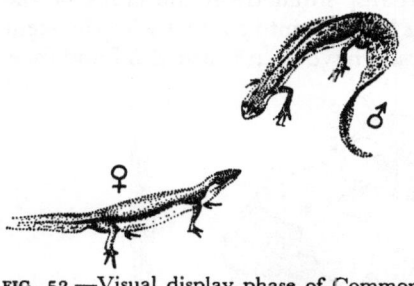

the Common Newt begins by posturing in front of the female, his crest raised, his side turned towards her (Fig. 53). He then executes a sudden leap, by which a strong water current is directed to the female, which often pushes her aside. He then

FIG. 53.—Visual display phase of Common Newt (*after Tinbergen and Ter Pelkwijk, 1938*)

faces her, bends his tail forward alongside the body, and by waving it sends a gentle water current which probably carries a chemical stimulant towards her (Fig. 54). If the female responds by walking towards the male, he turns round and crawls

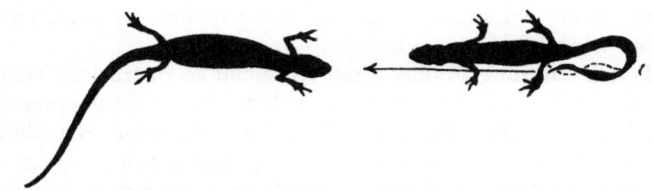

FIG. 54.—Male Common Newt sending a water current to the female
(*after Tinbergen and Ter Pelkwijk, 1938*)

away from her. After a while he stops, waits until she touches his tail, and then deposits a spermatophore, which the female takes up in her cloaca. Here again, experimental study is needed to test the obvious suggestion that the male's first movement is a visual display, the second sends out tactile stimuli, and the third provides chemical stimulation.

Visual releasers are relatively well known, though even here much more precise evidence is needed. The examples given so far have already shown that movement, colour, and shape may be involved. In some species the emphasis is on movement, as in the various courtship and threat displays of the Herring Gull. In other cases the emphasis is on colour, as in the red underside of the Stickleback, or the red patch on the Herring Gull's lower mandible. Usually, colour and movement are both involved, and then the movement is always well suited to show off the special structures that influence the reactor. Whether the movement is adapted to the structure or the structure to the movement, or both, is of course an evolutionary problem; I will return to it in Chapter VIII.

CONCLUSION

So far as our present knowledge goes, social co-operation seems to depend mainly on a system of releasers. The tendency of the actor to give these signals is innate, and the reactor's responses are likewise innate. Releasers seem always to be conspicuous, and relatively simple. This is significant, because we know from other work that the stimuli releasing innate behaviour are always simple 'sign stimuli'. It seems therefore as if the structures and behaviour elements acting as releasers are adapted to the task of providing sign stimuli. When releasers serve, in addition, the function of reproductive isolation, they are specific as well, that is, different from releasers in other species. This specificity cannot always be attained by one single releaser, but a sequence of releasers, each in itself not very specific, can in its entirety be highly specific.

Not all communication, however, is based on releasers; there are certain complications. As we have seen, many social animals respond to the species' social releasers only when provided by certain individuals, which they know personally. In such cases personal connexions, established through learning processes, confine the reactor's responses to signals from one or a few individuals only; they still respond to the releasers of the species, but only after they have narrowed their attention to particular members of the species.

The reactor's responses are sometimes immediate and simple movements. Often, however, they are internal responses; the signal in such cases changes the reactor's attitude and prepares it for a more complicated and variable activity.

We see therefore that a community functions as a result of properties of its members. Each member has the tendency to perform the signal movements releasing the 'correct' responses in the reactor; each member has specific capacities that render it sensitive to the species' signals. In this sense the community is determined by the individuals.

It is sometimes claimed by sociologists and philosophers that the individuals are determined by the demands of the community; at first glance this seems to be exactly contradictory to the above conclusion. Several discussions on this subject have failed to point out that each of the two conclusions is, in a way, true. The first conclusion is valid from the 'physiological' point of view, the second from an evolutionary point of view. When the individuals behave abnormally, the community of course suffers. Clearly in this sense the community is determined by its constituent individuals. However, only groups composed of 'capable' individuals survive, those composed of defective individuals do not, and hence cannot reproduce properly. In this way the result of co-operation of individuals is continually tested and checked, and thus the group determines ultimately, through its efficiency, the properties of the individuals. The argument can equally be applied to the individual and its constituent organs. Of course the individual is determined by its organs, in the sense that defective functioning of an organ endangers the life of the individual. The result of the organs' co-operation, the individual, is tested as a whole and only those with properly functioning organs survive. Thus the success of the individual determines, in the long run, its constituents.

CHAPTER VI

RELATIONS BETWEEN DIFFERENT SPECIES

IN the foregoing chapters we have seen that co-operation between individuals of the same species is often based on the releaser system. One individual, the actor, gives a signal to which the reactor responds. Such releaser-relationships are, however, by no means confined to relationships between animals of the same species; we know numerous cases of inter-specific co-operation based on a similar system of signals; some of these will now be discussed.

Two categories must be distinguished: (1) Many species have evolved devices the function of which is to release responses in individuals of another species. (2) Many species are specialized in the opposite direction: they do their utmost to avoid releasing a response in other species. More specifically: such species avoid releasing feeding responses in predators; they try not to draw the predators' attention.

THE RELEASE OF REACTIONS

This category is most impressively represented by the colours of flowers that depend for their pollination on insects. It is now known, mainly through the research of German workers, that many flowers are beautifully adapted to attract and guide the required pollinators.[24, 26, 35, 39, 40, 43, 79] Their main releaser is their colour. Since von Frisch, repudiating von Hess' claim that Honey Bees were colour-blind, showed that Honey Bees could distinguish very well between colours, the same has been shown for Bumblebees, flies, butterflies, and moths. The visits of many of these insects to flowers are guided mainly by colour. It is easy for instance to train Honey Bees to visit yellow or blue paper by presenting them with a sugar solution in a dish placed on such paper. In a critical test the sugar water is taken away, and an array of papers of various colours, and of a finely graded series of grey papers is offered. The bees then go straight to the colours for which they have been trained. It is possible to take the necessary precautions to make sure that they cannot react to either ultra-violet or infra-red, and this simple method is sufficient to show that bees can see colours.

The importance of the colours of flowers to bees has been studied in a number of cases. Knoll for instance noted that bees visiting the yellow flowers of *Helianthemum* alighted occasionally on other yellow flowers as well. When he robbed the Helianthemums of their yellow petals, leaving the rest of the

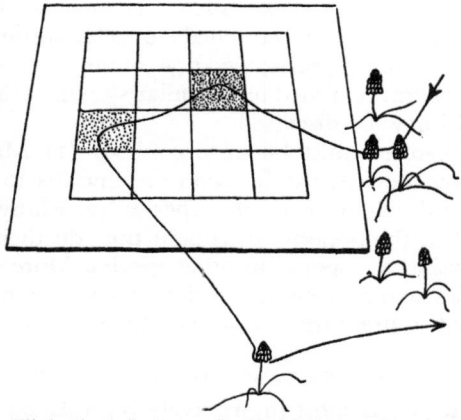

FIG. 55.—Flight line of a hover fly (*Bombylius*) to Grape Hyacinths and blue papers (*after Knoll, 1926*)

flowers intact with honey and pollen, the bees ignored these flowers. However, when he restored their appearance by attaching yellow paper petals to them, the bees came to them as before. Similar tests were done with a species of Hover Fly, *Bombylius*, and the blue flowers of the Grape-Hyacinth. When a chess-board of small paper squares of various colours and various shades of grey was erected among the Grape-Hyacinths the flies went to the blue squares but not to any of the other colours or greys (Fig. 55).

Many plants have coloured leaves around the flowers,

which, although not strictly belonging to the flowers, add greatly to their conspicuousness. *Salvia horminum*, a common garden annual, native to the Mediterranean countries, has a 'crown' of deep violet leaves, which are, as a matter of fact, much more conspicuous than the small, pale mauve flowers themselves. Bees in the Mediterranean countries react to the bright crown first, and then descend to the flowers. Knoll further observed that Honey Bees in Prague, where the plant was only found in the Botanical Gardens, did not know where to find the flowers at first, and, after having been attracted by

the crown, searched a long time among its leaves before they stumbled upon the flowers (Fig. 56).

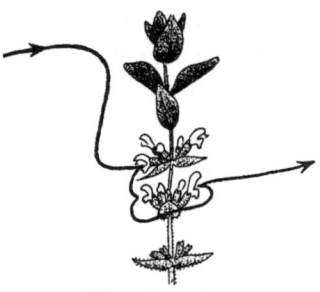

FIG. 56.—Flight line of a Honey Bee attracted by the violet 'crown' leaves of *Salvia horminum* (*after Knoll, 1926*)

An unexpected result appears when the chessboard test as described above is done with insects visiting the ordinary red Poppy. Bumblebees, for instance, while clearly attracted strongly by the Poppy's flowers, will not come to red papers presented near them. This is due to the fact that insects do not react to the Poppy's red colour. In fact, most insects are not sensitive to red, and see only black where we see red. Red is, to them, 'infra-yellow'. These insects react to quite another type of light: to ultra-violet, which is reflected by Poppies. Insects not only see ultra-violet light far beyond our visual limits, but they distinguish it as a colour different from any other colour. The red colour of the Poppy therefore seems to be no adaptation to insects, but a mere by-product, whereas its ultra-violet colour is of great importance to it. In our flora genuinely red flowers are exceedingly rare. Most 'red' flowers are in reality purple, or a mixture of red and blue, and it is to the bluish hues in such flowers that insects react.

Really red flowers occur in regions where pollinating birds occur; many of the American flowers that are visited by

Humming Birds for instance, are a flaming red. Plants in our regions show similar adaptations to birds: berries eaten by birds, and probably dependent on this for their germination, are often bright red.

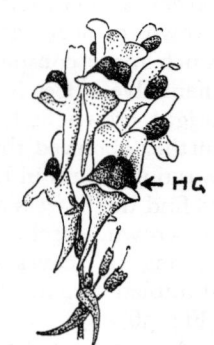

Many flowers show so-called 'honey-guides', patterns of dots or stripes arranged in such a way round the centre that they seem to lead to it. In some cases the guiding function of such honey-guides has been proved. The Toadflax, *Linaria vulgaris*, has a deep orange patch on the lower lip (Fig. 57), just below the entrance to the flower. The Humming-bird Hawk Moth, one of the species that have a tongue long enough to get at the honey deep down in the spur, aims its tongue tip exactly at this orange spot, and thus succeeds in finding the entrance. Reactions to a honey-guide of another type, that which has conspicuous

FIG. 57.—*Linaria vulgaris* and its orange honey guide (HG) (*after Knoll, 1926*)

stripes radiating from the flower's centre, have been observed by Knoll and Kugler in experiments with artificial flowers (Fig. 58).

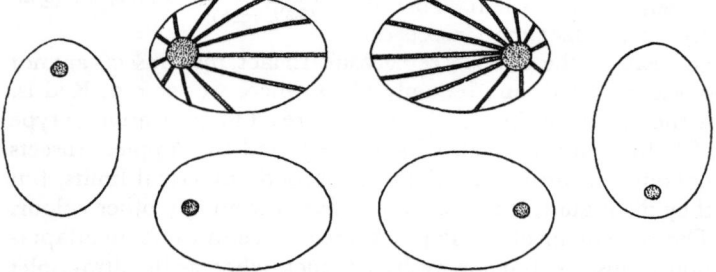

FIG. 58.—Artificial flowers with honey guide models. Visiting Humming-bird Hawk Moths were directed to the circular patches (*after Knoll, 1926*)

It is, of course, not only the flowers' colours that attract the insects; scents play their part as well. The way in which insects use the flowers' scent differs from one species to another.

PLATE 7

Eyed Hawk Moth at rest

Eyed Hawk Moth displaying 'eye'-spots on hind wings after being touched

Honey Bees and Bumblebees appear to be attracted by flower colour first. It is easy to lure them to coloured paper models. However, although they inspect them, they rarely alight on them, but turn away at about half an inch's distance. They do alight, however, when the paper flower is given the scent of the real thing. Scent to these insects merely provides a means of finally checking the flower's identity.

Some butterflies have been shown to react to scents in another way. They react to odours of various kinds, not by flying towards their source, but by reacting to coloured objects, mainly yellow and blue. The scent merely releases their visual responses, but does not direct them to the flowers. The strong scent emanated by many flowers that open at dusk has a different function again: it really attracts moths from a distance. I have seen a most impressive demonstration of this by flowers of the Honeysuckle, which I had concealed in a wooden box with a system of slits. The flowers were in the centre of the box, invisible from the outside, and through the slits the scent could disperse freely. At dusk, this set-up was visited by a number of Pine Hawk Moths, which reacted to the flowers from distances of up to ten yards. They zigzagged and circled round the box and soon found their way inside (Fig. 59). Knoll, who studied the visual responses of various Hawk Moths, found that they react also to the colours of the flowers, which they can distinguish even when it is already so dark that the human eye cannot see colours.

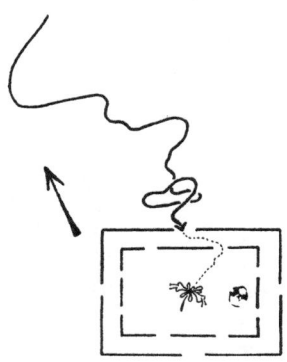

FIG. 59.—Flight line of a Pine Hawk Moth guided by the scent of concealed Honeysuckle flowers. Large arrow indicates direction of wind

As a last instance of the manifold relationships between flowers and insects, the trap-flowers must be mentioned. The best-known example in the British flora is *Arum maculatum*,

'Lords and Ladies'.[39] Each Arum-'flower' is really an in-florescence, enveloped in a large leaf, the spathe. The top of the inflorescence bears the 'club', which gives out an odour attracting numerous insects of various kinds. When these alight either on the club or on the inside of the spathe, they immediately fall down into the cavity of the flower, because the club and spathe are very slippery. The bottle-neck above the cavity has a circle of hairs, which stops the large insects and allows them to get away, but lets the small ones slip through. The slippery wall of the cavity and the equally slippery hairs pre-vent them from getting out. The only thing they can do is walk round and round over the inflorescence. During the first day the male flowers are closed, but the female flowers are open and ready to be pollinated. Since the visitors are only kept in captivity for one day, and since they fly from Arum to Arum, the chances are that several of them are already carrying Arum pollen. As soon as the female flowers are pollin-ated, the cells of the wall shrink, the wall loses its slippery nature, and all the insects can escape. Before this happens however the male flowers open, and the escaping insects all carry pollen with them. They deliver it in the Arum they next visit.

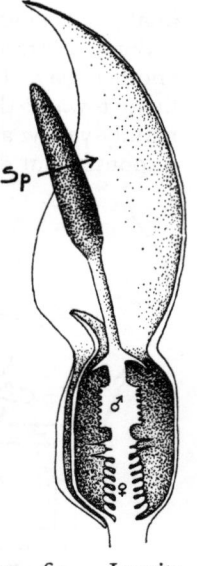

FIG. 60. — Longitu-dinal section of in-florescence of *Arum maculatum*, showing male and female flowers in the 'trap,' and spathe (Sp) (*after Knoll, 1926*)

Thus we see that plants have evolved numerous devices serving to attract and to guide pollinating insects. Many insects react to these devices innately. This is known for instance for the reactions of Bumblebees and Hawk Moths to colours, and it is probable in many other cases. Yet it is also known that Honey Bees, Bumblebees, and other insects learn to specialize now on this plant, now on another species. Exactly how innate responsiveness is mixed with various learning processes is not known in any particular case, and much work remains to be done in this field.

These interspecific relationships are mutual, just as are the intraspecific relationships. Both parties profit by the co-operation. There are however interspecific releasers which serve one-sided relationships, and, since these are of considerable interest in various respects, some of these will be discussed briefly.

The Lophiidae, a group of fishes of which *Lophius piscatorius*, the Angler Fish, occurs in the North Sea, have developed a signal that lures smaller fish to their doom. Lophius itself is beautifully camouflaged. On the top of its head it has a 'bait' which imitates, by size and movement, an animal of the size that releases feeding responses in small fish. When the latter come within reach of it, and before they can snap it up, Lophius opens its huge mouth and swallows its victim.[116] Lophius has thus developed a releaser which is adapted to the special sensitivity of its prey species, but the prey species certainly has not adapted itself to respond to Lophius.

A similar case is found in certain Orchids, such as the *Ophrys* species, the flowers of which resemble certain insects. The males of these insects react to the flowers, not to gather food, however, but in order to copulate with them; since their mating activity is a response to form and colour, and—so far as is known—to nothing else, they are misled by the flowers. In their attempts at copulation, they pollinate the flower.[4] Again, the adaptation is not mutual.

It is probable that the so-called deflection devices of some animals are similar one-sided releasers. In several fish the eye —the main structure which characterizes the appearance of the head—is concealed by such means as a dark band across it. At the opposite end of the body there is a conspicuous round dark spot. A tropical fish, *Chaetodon capistratus*, has the curious habit of swimming very slowly tail first; when disturbed by a predator it swims off quickly in the opposite direction.[13] It is possible that a predator, reacting to movement and 'eye' spot, will snap at the tail, and thus fail to get a firm hold. Cott mentions other examples of such deflecting devices; and although I believe that in several of these the eye spot on the tail may very well be a social releaser for intraspecific use, there can be little doubt that deflective devices exist. It would be highly interesting to do some experimental work on this problem.

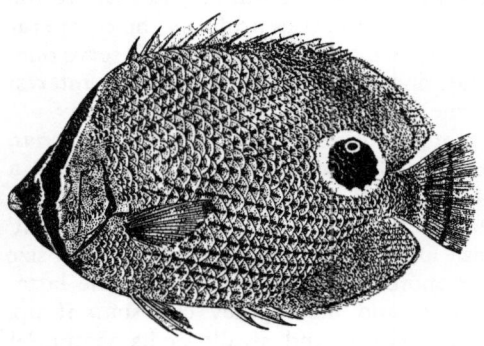

Another category of conspicuous colours are the so-called warning colours. Their function, like those of the flowers, is to release a response in animals of other species.

FIG. 61.—*Chaetodon capistratus* and its 'eye spot' (*after Cott, 1940*)

They do not however attract those animals, but repel them by releasing escape or withdrawal. They are aimed at predators. Here again we have to do with a one-sided relationship, for it is not to the predator's advantage to withdraw.

We must distinguish between two entirely different types within this category. In one type the colours have no influence on the predator until after the latter has learned that they mean harm. In the other type the predator is scared by a sudden display, and animals who use this type of defence use sheer 'bluff' for they are usually quite harmless, and edible as well. The name 'true warning colours' is often used for the first category, 'false warning colours' being those used in bluff.

Beautiful false warning colours are displayed by various butterflies and moths. The Eyed Hawk Moth, for instance, has a brightly coloured spot on the hind wings, much resembling a vertebrate's eye. The animal is nocturnal, and rests during the day. When at rest it is entirely camouflaged, and the hind wings are neatly concealed under the forewings. When it is touched, particularly when a sharp object such as a bird's bill hits it, it suddenly spreads its wings, thus exhibiting the hind wings, and waves them slowly back and forth (Plate 7).

PLATE 8

Larvae of the Cinnabar Moth, showing true warning coloration; predators have no innate avoidance reaction to them but learn to know the colour pattern as a sign of distastefulness

Experiments showed that birds were scared off by this display, and left the moth alone.[77] When the colours of the hind wings were brushed off however the display did not make the least impression on the birds, and the hapless moth was eaten forthwith. All over the world there are numerous species of insects showing such a sudden display of conspicuous colours. It has been demonstrated that their function depends entirely on the suddenness of the display; if you present such forms on a feeding tray with their warning colours plainly visible, they are eaten; it seems that most if not all such species are quite edible.

It would lead me too far to give a review of the various types of warning colours; moreover, they are treated in several books on this special topic.[13, 72] I will merely point out that many of them resemble eyes. This certainly is no accident; an eye is not only very conspicuous (so much so that cryptic animals have evolved numerous ways of concealing it) but many species, probably birds in particular, seem to be easily scared by the sight of a pair of eyes glaring at them from near by.

While much experimental work has been done on the functions of colours in flowers, on cryptic coloration, and on true warning colours, the experimental study of false warning colours has been much neglected. Almost all the evidence usually presented to support the idea of false warning colours is incidental and not very conclusive, and here again is a most attractive field of research practically untouched.

True warning colours exert their influence in a different way. They are never concealed, but are on display permanently. Common wasps offer a good example.[65] When a songbird such as a Redstart meets with a wasp for the first time in its life, it captures it. Sometimes, but that is relatively rare, the wasp will manage to sting the bird. The bird then lets go, and may show in various ways that the sting affected it rather unpleasantly; it may shake its head, and wipe its bill. Anyway it shows no further interest in the wasp. Usually however the wasp does not sting, it is killed before it can do so. Then it becomes evident that a wasp is distasteful: the bird does not finish it, and if it is eaten, it is often brought up again afterwards. Mostler has shown that most songbirds learn from one or a few such experiences to leave wasps alone. That they recognize such unpalatable insects by their colours is evident from the fact that from then on such a bird avoids not only wasps, but all similarly coloured insects. This type of coloration therefore does not act on the predator's innate responsiveness, but it serves to condition the predator to the colour as a sign of unpalatability. The same applies to the black and yellow pattern of the larva of the Cinnabar Moth (*Euchelia jacobaeae*). It has been shown by Windecker [117] that these too are tried by every young bird. They are distasteful because of some property of the skin and particularly of the hairs. In order to show this, Windecker mixed various parts of the caterpillars with Mealworms. If he mixed in the entrails, no bird objected. But as soon as the skin was mixed in, the birds refused the Mealworms with every sign of disgust after having once tasted them. The same investigator even managed to shave a great number of caterpillars, and mixed the hairs with the Mealworms. This too was enough to make the birds refuse.

Closely related to this type of coloration is mimicry. Mimics display the same type of colours as the species they imitate, though they are not themselves distasteful. Consequently they are refused by such predators as have had experience with the distasteful 'example'. This hypothesis, formulated long ago by Bates, has been most beautifully confirmed experimentally by Mostler. Hover Flies imitating wasps, bees, or Bumblebees were always eaten by inexperienced birds. As soon, however, as these birds had learned to avoid wasps, bees and Bumblebees, but not until then, they left the mimics alone.

There are also species that mimic each other mutually. Windecker showed that birds that have learned not to eat Cinnabar Moth larvae avoid wasps without further learning. In this way species can, as it were, shift part of the burden of the 'tax' which they have to pay for the 'education' of the predators on to the shoulders of another species. This type of mutual mimicry is known as Müllerian mimicry; Windecker's work is, so far as I know, the first experimental proof of its existence.

THE AVOIDANCE OF RELEASE

We will turn now to the second category of interspecific visual adaptations, those serving to avoid attracting attention. These comprise all types of camouflage. Camouflaged animals do their utmost not to present any stimulus that might release responses in predators. They have evolved the exact negatives of visual releasers. By a careful study of these negatives one gets further evidence, confirming that found by a study of visual releasers, of the kind of stimuli to which animals most readily respond. Whereas releasers are usually made more conspicuous by movement, camouflaged animals avoid movement as much as possible. Whereas releasers contrast in colour and shade with the environment, camouflaged animals adopt the colour of the environment. Whereas releasers offer simple patterns, camouflaged animals disrupt their outline, and have patterns which make it blend with the environment. Whereas the most specialized type of warning colours are 'eye' spots, camouflaged animals conceal their eyes. I must again refrain from describing many examples, referring the reader to Cott's book on adaptive coloration.[13]

There is rather substantial experimental evidence showing that such cryptic devices render their bearers less conspicuous not only to human eyes, but to those of their natural predators as well. A most convincing series of such tests has been carried out by Sumner [85, 86, 87] with *Gambusia*, a fish that can slowly change its colour, matching itself with the background. He presented fish that had adapted their colour, and some that had not yet had time to do so, in large tanks to various types of predators: Herons (hunting them from above), Penguins (birds hunting under water), and predatory fish—and in all cases he found that the more conspicuous fish were captured in far greater numbers than the camouflaged fish. Dice [19] offered mice of various shades, some of them matching the ground better than others, to Owls, and found that the Owls actually captured the less well camouflaged mice first. These, and most other experiments in this field, deal with the general resemblance of the colour of the animal to that of the background. Much work still remains to be done on the other aspects of camouflage: disruption of outline, hiding of the eyes, countershading, &c.

This review, short though it is, may be sufficient to show that releasers are not only used in social intercourse, establishing relations between animals of the same species, but also that interspecific relationships are often based upon them. They always serve to release behaviour in the reactor that is useful to the actor. Their main characteristics, conspicuousness and simplicity, are found in both intraspecific and interspecific kinds. On the contrary, specificity is found in only some of the interspecific releasers; in warning colours of various types, for instance, there does not seem to be any need for it.

CHAPTER VII

THE GROWTH OF SOCIAL
ORGANIZATIONS

DIFFERENTIATION AND INTEGRATION

THE relationship between an animal, for example a bird, and its young begins in essentially the same way as that between an individual and one of its organs. At the beginning the young is but an egg-cell in the body of the mother, one cell in an organ of the mother, the ovary. As soon as the egg has been fertilized, it begins to cleave and to differentiate. Through a number of complicated processes the mother's body supplies food, and forms supporting and protecting structures, and in this way the egg-cell becomes a more or less isolated whole, an egg. When the egg leaves the body of the mother, it becomes much less dependent on the mother than it was before: food and oxygen are no longer provided by the mother. It is not completely independent, however: the mother has to brood the egg. Differentiation goes on; certain groups of cells form the skin, others the gut, others again the brain, and so on.

When the egg hatches, the relationship between it and the mother changes abruptly. Brooding is still necessary it is true, at least at the beginning, but the eggs are no longer shifted, and new relationships such as feeding and removing the faeces are established. Further, the young bird begins to react to call-and warning notes. These new relationships are no less real, nor less vital than the old ones, although they are often less easy to detect. Apart from small changes, they function until the young one gains complete independence. In some species this is brought about by loss of interest, sometimes on the part of the parents, sometimes on the part of the young; usually it is mutual. Often the parents take the initiative by forcibly chasing the young away; this may be seen when the parents are beginning a new reproductive cycle. In other cases the bond between parent and young gradually changes into one between social colleagues; the family becomes a flock.

This type of community therefore begins as a relationship between an individual and one of its organs, and it changes gradually into one between individuals. Typical of this development is a growing independence of the organ, and its growing differentiation. A community has evolved out of one individual through continuing differentiation of its organs. Such a differentiation may lead to extremely complicated societies, such as the 'states' of social insects. I will discuss some examples, beginning with rather simple relationships between mother and offspring, and then proceeding to the more complicated types.

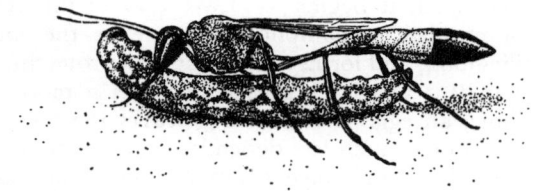

FIG. 62.—The Digger Wasp *Ammophila adriaansei* with prey
(*after Baerends, 1941*)

Most insect states take their origin in a fertilized female. Many insects abandon their eggs as soon as they are laid, and the 'community' never passes beyond the relationship between individual and organ. Many bees and wasps, however, continue to care for the eggs after they are laid, and even for the larvae after they have hatched. Some solitary wasps for instance, such as *Ammophila adriaansei* (Fig. 62), not only provide a paralysed prey as food for the larva, as most digger wasps do, but bring new food when the larva has eaten the first store. When the larva begins to spin a cocoon the mother leaves it; she dies long before any of her offspring hatches.

Among the 'solitary' bees we find some species that have reached a higher level of social organization than these digger wasps. A burrowing bee, for instance, *Halictus quadricinctus*, not only provides the eggs with a store of honey and pollen, but stays in the burrow until the larvae have hatched; she associates with her offspring. The offspring do not leave the burrow but expand it, lay their eggs in it and take care of the brood. Each brings food for its own larvae as well as for other larvae in the hole. The last generation to hatch in the autumn, however, is not so socially disposed; it leaves the nest, and the individuals scatter. They hibernate on their own, and those that survive found a new 'family' next spring.

The Bumblebees have evolved a further very important step in social organization. A Bumblebee-community is again founded by a female, a 'queen'. This queen comes into close contact with her offspring; she occasionally opens the cells in which the larvae are growing and replenishes their food store. The first larvae all develop into females. These early females have underdeveloped ovaries and are sterile; they are 'workers'. From that moment the queen becomes more or less an egg-laying machine; the workers do all the other work: they build new cells, fly out to gather food, and feed the queen and her offspring. In a community of Bumblebees there is, therefore, division of labour between the component individuals. In late summer the eggs produce more completely developed females, and also males. These mate, and in autumn the large family disintegrates. All except the newly fertilized females die. These prospective queens hibernate, sometimes on their own, sometimes in groups in the old nests, but next spring each of them starts the long search for a new burrow in which a new community can be founded.

The social bees, of which the Honey Bee is the best known, proceed still further. First, the division of labour is carried to an extreme degree.[75] As in Bumblebees, there is a queen, sterile females or workers, and males. The workers have a variety of tasks in the community. Some of them collect honey, others pollen. Others again do nothing except build new combs, others again specialize in parental duties and care of the brood. This division of labour is a matter of age: each worker holds these 'offices' in successive periods of its life. Shortly after a worker has left its cell, it begins to sweep and clean cells from which workers have recently emerged. Only after a cell has thus been cleaned is the queen willing to provide it with a new egg. When the worker has stuck to this job for about three days, she begins to feed the larvae, and particularly the older ones. To this purpose she collects honey and pollen from the stores. After another three days have elapsed, she starts feeding the younger larvae as well. These get a different food; apart from honey and pollen they receive a kind of 'milk', an easily digestible food secreted by special glands in the worker's head. Workers of this age also venture into the open for the first time in their lives; they make short reconnaissance flights, but without as yet gathering honey or pollen. At the age of ten days, the worker abandons its former work; the brood no longer interests it and it embarks upon various household duties, such as taking over the honey from incoming foragers, depositing the honey in the cells or feeding it to other bees; it stamps the pollen brought in by the foragers tight into the pollen cells, it builds new cells, using the wax secreted by its wax gland, and it carries away dead bees and rubbish. On its twentieth day it becomes a guard: it posts itself at the hive's entrance and inspects every arriving bee. Twenty to thirty of these guards are in office simultaneously; they attack and drive off each intruder. They do not remain guards for long, however; soon they become foragers, flying out into the country and collecting honey and pollen; this they do until they die. Among the foragers there is a further division of labour; some of them are 'scouts'; they find new food plants when the food from the species of plant in use at the moment is running short, and by their 'dances' convey the kind, direction, and distance of any food source they happen to find.

A Honey Bees' community differs further from a Bumblebees' in that it does not disintegrate in the autumn. Unless it is disturbed, it remains in existence year after year. The community thus exists longer than any of its constituent individuals, and that is why such communities are called 'states'. A new state is not founded by a solitary queen but by a 'swarm', composed of a queen with workers of all classes. The original state, possessing one queen, divides just before a new queen is born; the old queen leaves with the swarm to a new site. Later in the season more swarms may leave the hive, each headed by a young queen. New states therefore develop by a process reminding us of cell-division. In both cases the daughter-organisms, after gaining independence, have to grow up by their own labours.

In ants, all species of which are social, a new colony is founded in one of various ways. In many species fertilized females settle down and begin to lay eggs from which the first workers of the new colony will emerge. In other species the queens cannot live on their own and have to acquire the help of a number of workers. In some of these species the queen leaves the nest together with a crowd of followers; in other species the queen enters into an existing nest of her own species, and forces the original queen to leave. The queens of some of these species may enter a nest of another species, kill all the adult ants and adopt the brood; in this way the curious phenomenon of 'slavery' originates. In other species again one nest may have many queens; from time to time one of them departs with a group of workers and founds a new colony.

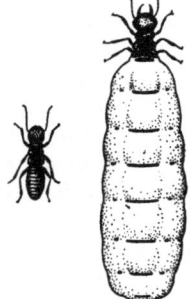

FIG. 63. — Termite 'king' (*left*) and 'queen' (*right*)

The communities of termites, though showing many amazing convergencies with those of ants, are derived not from a mother-family, but from a pair and its offspring. Males and females play equal parts: there is a 'royal couple' (Fig. 63); among the workers there are males and females in equal numbers. Both are sterile. From time to time fertile males and females which are winged hatch. They leave the burrow together in a huge swarm. After swarming they lose their wings, and pairs are formed on the ground, where the females attract the males by means of a special scent-gland. The members of such a pair are not yet sexually mature; they do not copulate but dig a hole first, the beginning of the termite burrow, which will house their future offspring. After some time the pair copulates and eggs are produced. The larvae of termites are not as helpless as are those of bees or ants, and take part in many activities of the colony. They gradually develop into adult workers, diverging into the various 'castes'.

Recently, another way of founding new colonies has been described similar to one of the methods of some ants mentioned above. Grassé and Noirot [28] have observed how dense columns emerged from a nest and travelled a certain distance to found new nests. One of the columns included the royal couple. All the castes were represented in the columns, even winged individuals. In the groups which did not contain the royal couple 'substitution reproductives' developed through 'neoteny', larvae becoming sexually mature before their time. Grassé has named this method of splitting up of a society into equivalent daughter societies 'sociotomy'.

Apart from sociotomy, which strictly speaking is not a mode of origin of a new society, all the instances discussed so far show the arising of a community through differentiation of the mother-egg relationship; in the termites the father is taken up into the state as well. This type of origin could therefore be called 'growth' or 'differentiation'.

Not all types of social organization start in this way, however. Many communities are formed by independent individuals coming together and associating, thereby losing their independence.

This happens for instance when male and female form a pair, and when Starlings flock together. Ties are formed which did not exist before. This type of development of a community could be called 'construction', or 'integration'. The two processes, differentiation and integration, move in opposite directions; in the former, total dependence of one of the partners develops into a state of mutual co-operation; in the latter, mutual co-operation takes the place of total independence.

THE ESTABLISHMENT OF SOCIAL TIES

How, in these two types, does co-operation originate, how are social relationships established? We have seen above that co-operation is ensured by a system of innate activities in the actor, and of (usually innate) responsiveness to the actor's behaviour in the reactor. The satisfactory functioning of these behaviour elements is as a rule ensured by 'preparedness-in-advance'. A bird develops the tendency to incubate eggs some time before it lays eggs. The readiness to feed young is there before the eggs have hatched. Such tendencies normally remain dormant until the outside objects to which they react appear and provide the releasing stimuli. Under abnormal conditions, and even sometimes under normal conditions, they lead to behaviour in the absence of the adequate releasing objects. Many birds, for instance, begin to sit in the nest before eggs are laid. What matures in the bird is not a mere readiness to respond to eggs, but an urge which may lead to overt behaviour even when the eggs are absent. We all know comparable behaviour in the human female: a childless woman often provides herself with a substitute with which to satisfy her maternal drive: either an adopted child or a pet. Many childless women develop a curious ambivalent attitude towards their own husbands and make them play the double rôle of mate and child.

In the integration-type of founding a community social co-operation is established in the same way. The potentialities of acting and reacting to the partner are usually prepared beforehand. Neither a Grayling nor a Stickleback need learn how to recognize or how to react to their social or sex partners.

FURTHER DEVELOPMENTS

With the establishment of a tie between two or more individuals the development of the relationship is not always completed. Various subsequent changes may occur, which have now to be discussed.

In some cases we may notice a gradual change, such as an increase or a decrease in social activities. Such a change has been studied in the parental activities of male Sticklebacks. One of these is 'fanning', a movement by which the male, by special movements of its fins, sends a water current into the nest, thus carrying oxygen towards the eggs and removing carbon dioxide. When the eggs are still young, only a minor part of the male's time is taken in fanning. Later, the eggs demand an increasing amount of oxygen, and naturally give off increasing amounts of carbon dioxide. The male meets these needs by spending progressively more time in fanning. This increase in his activities is partly due to an increase in the intensity of the stimuli provided by the eggs: when eggs of eight days old are placed in a nest containing eggs three days old, the male reacts by a noticeable increase of fanning. The normal increase in fanning activity in the course of the eggs' development is, however, partly due to an internal change in the male: by replacing the eggs in the nest by fresh eggs at various stages of the male's cycle, the male's fanning, while always dropping somewhat, does not decrease to the level of the first day. The later in the cycle we give the fresh eggs, the more intense is the male's reaction to fresh eggs.

In a similar way the incubation drive in a sitting bird grows with time. This happens also when the eggs are killed or are infertile.

Gradual changes of a more complicated type have been observed in the pair formation processes of birds and fish. A good description has been given by Verwey [113] of the pair formation of Blue Herons. The birds live solitarily during the winter and return to the breeding colonies in spring. The males arrive first, and take up a position on an old nest of the previous year, or on a place where they are going to build a new nest. Here each of them utters the 'song', a harsh, monosyllabic call, not very pleasant to the human ear, but attractive to the females. When a female arrives, it settles on a branch near the male of her choice. The male begins to court at once, but when the female responds by approaching, he wards her off, and a skirmish or even a furious fight may ensue. When the female flies off, the male at once resumes his frantic calling, and then she may turn and return to him. This again may evoke hostile reactions, but gradually the aggressiveness subsides, the birds begin to tolerate each other, and eventually mate. It is clear that the male, and presumably the female as well, reacted to the partner in two ways: by a sexual response, drawing them together for the purpose of mating, and by an aggressive response, possibly mingled with fear, or the tendency to escape. Gradually the sex drive overcomes the hostile tendencies. This change in the relative strength of the various drives involved may be due in part to a learning process, the birds getting individually used to each other. In part it may be due to a growing sex drive under the influence of the repeated and prolonged sexual stimulation from the partner. That growth of the sex drive plays a part is indicated by the fact that skirmishes are rare, or short, in pairs that form later in the season. Males that have been waiting for a mate for a fortnight are so strongly motivated sexually by the time a female joins them that they may accept her almost at once.

In the Three-spined Stickleback, which as we have seen mates only for the purpose of fertilization, and in which there is no question of individual attachment between sex partners, the change from initially hostile to purely sexual behaviour depends entirely on the sex drive over-riding the hostility.[106] The male's first reaction to an approaching female, the zigzag dance, is an expression of two drives. Each zigzag begins with a move away from the female. This part is the beginning of a purely sexual response: swimming-to-the-nest, where alone the male can fertilize eggs. This is clear from the facts that (1) the 'zigs' may, under certain conditions, develop into complete 'leading' when the male swims the whole way to the nest, and (2) the zigs are most pronounced when the sex drive is strongest. The 'zags' are movements towards the female. In extreme cases, they develop into real attack; this happens when the attack drive can be shown to be exceptionally high. The female's reaction towards the male's zigzag dance provides a strong stimulus acting upon the male's sex drive. When she turns towards him he stops his zigzagging at once and swims towards the nest. The entire chain of his activities following this—swimming to the nest, showing the nest entrance, quivering, and fertilization—is predominantly sexual. Here the male's mixed behaviour, the zigzag dance, changes into purely sexual behaviour merely because the female, as a reaction to his zigzagging, provides a new, sexual stimulus, which tips the balance of the male's behaviour, switching him over into purely sexual behaviour.

After the female has spawned and the male has fertilized the eggs, his behaviour at once reverts to attack; he drives the female off. This is due to two changes: first, the male's sex drive, after sperm ejaculation, drops abruptly and thus no longer competes with his attack drive, which is as strong as ever; second, the female, having laid her eggs, has no longer a swollen abdomen and thus fails to provide one of the releasers evoking the male's sexual response. She now offers mainly attack-releasing stimuli.

Many changes in social structure occur as a consequence of learning processes. These often make the ties more specific; the reactor, who began by responding to stimuli given by any actor, begins to confine his responses to stimuli given by a certain individual. This is usually achieved by conditioning, a relatively simple type of learning. Parent Herring Gulls become conditioned to their young in the course of a few days, and from then on confine all their parental activities to them, becoming indifferent, or even hostile, towards other Herring Gulls' chicks. As described in Chapters I, II and III such personal relationships are now known in many birds; so far as we know they play an even more important part in many mammals. It is evident that such personal relationships cannot exist when the animals are merely reacting to sign stimuli characteristic of the whole species; conditioning obviously makes them react to many more stimuli, which enable them to distinguish between individuals. This power of discrimination is often amazingly acute; many birds for instance recognize their mates, or their chicks, or their social companions, at a glance, when the ablest human observer fails, or at the best can only just distinguish between them. The human failure is partly due to lack of training; when one associates closely with a group of, say, geese, or sheep, one learns to know every individual. However, I have never known a man to become as good at it as the animals themselves. It may be that each species shows the best achievements when distinguishing between individuals of its own species. However this may be, the promptness of responses limited to certain individuals shows that stimuli of a very subtle nature must play a part, in strong contrast to the sign stimuli to which an animal responds innately.

There are some scattered observations in the literature which tell us something about the nature of these stimuli. We know for instance that Terns and Gulls recognize their mates both by voice and by sight. Recognition by voice can easily be observed in the breeding colonies of these species. An incubating bird dozes off every now and then. It is fascinating to watch such a dozing bird from a hide. In a colony of Common Terns (*Sterna hirundo*) for instance, many birds fly back and forth. Both parents take their turn in incubation, and each sits for about an hour. The sitting bird may be alone for much of this time. It often does not pay the slightest attention to the passing birds, most of which are calling as they fly by. It reacts at once however when its mate arrives, and, since its eyes were closed, it must have reacted to the partner's calls. It is not difficult to watch such an immediate reaction to the mate's voice several times a day.[92] Such responses are often remarkably acute: the mate's call may be faint and distant and scarcely audible among the noise caused by so many other birds, yet the sleeping bird wakes up in a flash.

However, a bird may also recognize its partner among numerous strangers when it is silent. In the Herring Gull, a species I have watched more intensively than Common Terns, I have seen proof of recognition among partners when they were some twenty-five yards apart and when I was certain voice played no part. This visual recognition has probably to do with facial expression, dependent on the proportions of various parts of the head, as in humans. Human observers can detect differences in facial expression in animals easily, and there is one interesting observation by Heinroth indicating that a bird may fail to recognize its mate when the latter's face is concealed: he once saw a Swan in the Berlin Zoo attack its mate while she was 'up-ending' with the head under water. He at once stopped his attack when she showed her head above the water. Lorenz has made similar observations in his Greylag Geese.[55]

Experimentation on this problem is difficult, probably because the animals react to so many details at once that a change of some of these details, while clearly not remaining unnoticed by the bird, nevertheless leaves enough unaltered to make recognition still possible. We once changed the colour of Herring Gull chicks in order to confuse the parents. When we blackened a chick by rubbing it with soot, the parents looked startled, but they adopted the chick nevertheless, possibly because they recognized its voice. The same happened when we changed the pattern of dark blotches on a chick's head. We never continued this work beyond some preliminary tests, however. Although tests of this type take much time, they would be worth doing.

It may be mentioned here that some Penguins have developed another type of parent-young-relationship.[74] The young of Adélie Penguins and other species unite into large flocks, and are said to be fed by the parents indiscriminately. This 'crèche'—system (Plate 5, lower fig.) has been regarded by some authors as an adaptation to low temperatures, since huddling together reduces loss of heat. Some authors claim that the Sandwich Tern has a similar crèche system. My own experience suggests that, though many chicks may aggregate in flocks, each of them is usually fed by its own parents, which recognize their own young individually.

Relations between individuals may also become more specific by a process which seems to be quite distinct from conditioning. Heinroth reported the following remarkable experience. He had hatched a number of goslings in an incubator. When they had hatched, he took them out and carried them to a pair of geese which had just hatched young of their own. To his astonishment the incubator-hatched goslings did not associate with these geese, but ran back to him every time he put them with the geese. Clearly they considered him the 'mother goose', and did not recognize their own species at all. He found out that this did not happen when the goslings had no chance of seeing him before they were presented to the old geese. Later, Lorenz had the same experience with goslings, and also with various species of duck. Apparently the young of such birds have to learn what their own species looks like, and they learn this in a very short time. With geese, this seems to be a matter of seconds. This curious process is called 'imprinting'; its characteristics are the short time needed for it, and, as Lorenz claims, the fact that it cannot be reversed or undone. Once a gosling is attached to a human being, it is impossible to make it consider itself a goose, however long it is forced to live with geese. However, here the evidence is still conflicting, and further research is needed.

This, of course, does not mean that the goslings are born without any 'knowledge' whatsoever of what their social companions look like, or, in other words, that they would not react innately to any stimulus provided by the parents at all. Since they attach themselves to human substitutes or to other animal species, and not, as a rule, to plants or inanimate objects (an exception was a Blue Snow Gander on the New Grounds, which apparently was imprinted to its kennel-type nesting box [81]), these substitutes must provide some stimuli to which the goslings react. One of these stimuli is movement; Lorenz and I once showed this in some tests we did with an incubator-hatched Egyptian Goose. We took it up in a closed box to a bare room. When we were sitting motionless each in a corner, we released the chick, which then did not come to either of us, but stood helplessly in the centre of the room, calling frantically. When a cushion was pulled across the room, it ran after it, but abandoned it again as soon as it stopped moving. Fabricius [21] did more extensive experiments of this kind with newly hatched Tufted Duck and other species. He found that movement and calls were the stimuli provided by the parent. Movement however did not act as such, but it was necessary that parts of the body, the limbs, moved in relation to the rest of the body. Movement was of such importance that the ducklings followed a waving hand most readily, but did not pay the slightest attention to a motionless mounted Tufted Duck. The sensitive period, during which imprinting was possible, lasted till about thirty-six hours after hatching, but even chicks brought in touch with foster parents after only eighteen hours' isolation failed to become completely imprinted.

A similar process was discovered by Noble in Cichlid fish. In the Jewel Fish, *Hemichromis bimaculatus*, the parents become imprinted to their young. As described in Chapter III, he could imprint an inexperienced pair to young of another species by exchanging eggs; a pair thus treated was spoiled for future breeding since they never accepted their own young.

This imprinting does not lead to individual recognition, which in the case of geese and ducks is acquired later and more slowly. In the Cichlids, parents do not recognize individual young—considering that one brood may contain several hundreds, this would demand too much.

Certainly imprinting deserves further careful study; it is not only interesting to find out to what stimuli the birds are reacting just after hatching, but also what effect imprinting really has, and why, in many cases, it cannot be forgotten or changed.

Further study of the behaviour of human-imprinted geese showed the curious intermingling of acquired with innate responses. Goslings following their human parent lag farther behind than wild goslings do when following their parents. This is determined by the angle subtended by the parent. They keep at a distance from which the man subtends the same angle as a goose, and since a man is so much larger, this increases the distance. When the human parent swims, the part showing above the water is much lower than a goose, and correspondingly the goslings come very close. When he sinks his head slowly under water, the goslings come closer and closer and finally crawl on to his head.

Lorenz's goslings still associated with him after they had begun to fly. Although he was not able to join them in their flights, they took off now and then for trips over the surrounding country, partly satisfied by each others' presence. Every now and then they would alight, and then walk towards Lorenz as quickly as they could do. It was discovered by accident why they did not alight as near to him as wild geese would to their parents. Lorenz used to bicycle along the road to keep pace with the flying geese. Once, looking up into the air watching his birds, he fell into the grass bordering the road. Immediately the geese alighted near him. After that he could always induce them to alight by running fast and then falling down with his arms spread, thus imitating the movements of an alighting goose. The reaction to this movement must have been innate in the geese; they expected this releaser even in the foster parent to whom they were imprinted.

The process of conditioning to the parents, whether in the more specialized form of imprinting or not, has another interesting aspect. When a young Jackdaw is reared by hand, it becomes attached to its human foster parent. It keeps his company, and wants food from him. When such a human-grown Jackdaw begins to fly, human company does not satisfy it any more, and it associates with birds in all activities that involve flying. When there are wild Jackdaws or Crows around these become its flying companions. When it reaches sexual maturity, it shows, in spite of its long association with Jackdaws, that its education has left traces: its courtship is directed at human beings. When, later in the season, its parental instinct awakes, it selects young Jackdaws again, and not human babies. The object of its attention therefore depends on what instinct is aroused. One Jackdaw, famous among ornithologists, Prof. Lorenz's 'Jock',[57] treated her foster-father as her parent, Hooded Crows as social foraging-companions, a young girl as her husband, and a young Jackdaw as her baby.

These curious relationships, developing under abnormal conditions, reveal something about the processes responsible for social organization. They show that such animals see their environment, more particularly a fellow-member of their own species, in a peculiarly particulate way. They do not learn, as we would assume: 'That is what my kind looks like', and then direct all their social activities to their own species, but the different parts of their behaviour pattern are responses to different stimuli from the companion. Since all these stimuli are really provided by every member of the species, the kaleidoscopic nature of this bundle of reactions does not become evident; under abnormal conditions it is revealed.

CONCLUSION

Although, as the reader will have perceived, our knowledge of the development of social structures is still patchy, what little we know shows one thing clearly: many animal communities depend on the functioning of remarkably few and simple relationships. Whether a community differentiates from a simple body-organ relationship, or is constructed by two independent individuals joining into an organization, the relations between the individuals, based on the releaser-system, begin to function as soon as they are needed, or even before. The potentialities are always ready in advance. After they have started, various changes may occur. These may be due to changes in the intensity of the underlying drives, or to learning processes. Of these, imprinting conditions an animal to its own species, and other learning processes may condition it to individual companions.

REGULATION

When studying the way in which a community is organized, one is often struck by the many parallels that can be drawn between it and an individual. Both are composed of constituent parts; the individual is composed of organs, the community of individuals. In both, there is division of labour between the component parts. In both, the parts co-operate for the benefit of the whole, and through it for their own benefit. The constituent partners give and receive. Thus they lose part of their 'sovereignty' as well as part of their capacity to lead a life in isolation. The loss of sovereignty can go so far that parts give their own life for the benefit of the whole. There is constant loss of skin cells in the individual; a lizard's tail is left for the predator for the benefit of the rest of the lizard, so that this rest can live and reproduce. A mother duck defends her chicks even at the cost of her own life. The benefits that the parts derive from the whole is obvious in the individual; an isolated muscle cannot live long. But neither can an isolated worker of the Honey Bee nor an isolated polyp of a Siphonophore colony. Even in cases where individuals can live in isolation, they lose the manifold benefits they receive when living in the flock, as shown in Chapter III. Loss of the capacity to live outside the community is more striking in the organs of an individual—which has derived its name from it—than in the community; yet the difference is one of degree only. There are individuals which can very well be divided into parts without fatal consequences; tape worms, Planarias and sea anemones are not 'undividable'.

Comparison of individual and community, leading to the idea of the community as a 'superorganism', is of great use to the sociologist. Of course it must not be carried too far; organism and community cannot be identified; yet it helps one realize that in both cases one has to do with a 'going concern', presenting problems of organization and co-operation. The main difference between individual and community is one of level of integration; in a community integration has been carried one step beyond the individual.

So far, we have been studying the normal functioning of communities. What happens if something abnormal occurs? It is well known that an individual can in some cases respond to abnormal conditions in an adaptive way. Not only can it meet and withstand the numerous destructive influences to which it is exposed under normal conditions, but it can also cope with certain emergencies. This it does by so-called regulations. When a part of the body is damaged, the wound, if not too extensive, heals. When this cannot be done, another part may take over its function. Various examples of this remarkable capacity have been given by E. S. Russell.[78] Such regulations are, in a sense, nothing but an extension of the normal activities.

When part of the body of an individual regenerates after having been damaged, this is done by cell groups which by a kind of regression return to something like the state of embryonic cells; the cycle of growth begins anew. When the functions of a lost part of the body are taken over by another part, something different happens: the latter part extends its normal activities and realizes potentialities which would never have been realized under normal conditions.

Similar regulations do indeed occur in the community. Here, too, the constituent individuals may regress and start a new cycle. In other cases, abnormal conditions may make individuals do things which they would not have done otherwise; they may take over the task of lost individuals. To this purpose they have a number of mechanisms in reserve which come into play only in an emergency.

When birds lose their brood, they often start a new one. Instead of continuing their development as if nothing had happened, and proceeding from the incubation stage to that of care for the young—without contributing anything to the species—they undergo a profound change. Their testes and ovaries begin to develop sex cells again, their courtship begins anew, they copulate, build a nest, and lay eggs. Regulatory capacities of this kind are not the same in all species, but most, if not all, birds have them.

A most fascinating example of such 'regeneration' has been discovered by Roesch[75] in Honey Bees. As described in Chapter VI there is a rigid division of labour among the various age groups in a bee-community. When one of these age groups is artificially eliminated, the other groups take over the duties of this group and thus save the superorganism. When for instance all pollen- and honey-foragers are taken away—usually bees of twenty days or over—young bees of scarcely six days old, who normally feed the larvae, fly out and become foragers. If all building workers are taken away —those between eighteen and twenty days old—their task is taken on by older bees, who had already been builders before but who had gone on to the stage of forager. To this end they not only change their behaviour, but also regenerate the wax glands. The mechanisms of these regulations are not known.

In birds of prey male and female have different tasks in feeding the young. The male hunts, while the female guards the brood. The prey brought by the male is passed to the female, who then tears it up, and feeds small bits to the young. Not until the young are half-grown are they able to master prey entirely by themselves. This division of labour is so rigid that the brood is usually lost if the female dies during this period. In some cases, however, it has been observed that a male, after some delay, begins to feed the young after the manner of the female; a type of behaviour never observed in males of these species under normal conditions.[90]

Minor regulations, brought about by behaviour patterns that do not normally appear, though always kept in reserve, can frequently be observed. In Chapter III it was described how a male Ringed Plover may drive the female to the nest if for some reason she stays away from it. Once I watched a male Lapwing trying to drive its own fully fledged young away from a cat when they did not respond to his alarm-calls. Many songbirds have special reactions to non-gaping young which make them gape if the normal stimulus has no effect.

It is of course difficult to draw the line between normal and abnormal; 'normal' in this context means nothing else than 'often observed', and abnormal means rare; there are inter-mediates of all kinds. But the same applies to all 'regulations' whether of the individual or of the community. This shows once more that all regulations are but extensions of the normal life processes. In this respect it is well to remember that the normal life processes are neither more nor less mysterious than the regulatory processes; the latter do not present a problem quite apart from the former. When it is recognized that normal co-operation can be analysed, it is clear that we can apply the same methods to regulations; mechanisms kept in reserve need not be fundamentally different from mechanisms used daily.

CHAPTER VIII

EVOLUTIONARY ASPECTS OF SOCIAL ORGANIZATION

THE COMPARATIVE METHOD

WE have no documents about the history of social organization; fossils tell us little about the behaviour of animals of the past. We cannot therefore study the history of social organization directly. Yet it is possible to learn something about it by comparison of the social organizations of present-day species. Comparison is widely used for this purpose in morphology. Before we apply it to social behaviour, let me recapitulate how it is applied in morphology.

The first step in comparison is to study similarity and diversity, and to arrange animal species in groups according to these criteria, placing similar animals together in a group, similar groups together in a larger group, and so on. Similarity is taken as proof of affinity. In assessing similarity, one difficulty is encountered: the resemblance between species or groups may be superficial, and may 'fake' affinity. For instance, at first sight whales and fish are very much alike. On closer inspection, this similarity appears to be based on both having a streamlined torpedo shape—a character which happens to impress us unduly. In a great number of other respects, however, they differ greatly: in skeleton, skin, nasal cavity, reproduction, &c. In all these respects whales are much more like mammals than like fish, and thus, essentially by weight of sheer majority of characters, the whales are taken to be more related to mammals than to fish. Palaeontology confirms this conclusion.

Whales resemble fish because they have adapted themselves to a similar environment, which made them develop the similar adaptive streamlined shape. This phenomenon of similar adaptation has occurred in many animals, and is known as convergence. Convergences can be traced in every life process, in growth leading to 'structure' and in growth leading to 'function'—'structure' and 'function' of course being two aspects of one thing: functioning structure. Convergences can be traced in the animal as a whole, such as in whales and fish, in bats and birds, in gulls and fulmars. They can also be traced in organs, such as the hands, adapted to burrowing, of Moles and Mole-Crickets, the touch receptors of insects and mammals, &c. In assessing affinity, convergences are discarded, and true resemblance, or homology, is the only criterion.

When animals of one group are compared, a general pattern is found to be common to them all. Species, or groups of species, that differ in many respects from this pattern, are then considered to have diverged from the general scheme in these respects. Those that conform most to the general pattern are considered to be, in these respects, more similar to the original ancestors. Thus whales and bats are specialized as regards their adaptations to the medium, while, in other respects, they are just ordinary mammals.

Different species of a group, or different small groups of a larger group, may have evolved in the same direction, but one may have gone further than the other. This often makes it possible to detect trends of evolution by arranging species in a series, ranging from the most specialized species to the less specialized ones through intermediates. This procedure has many pitfalls; it must always be remembered that within a group we can rarely consider one animal as a whole as less specialized than another; in some respects it may be less specialized—in others more.

COMPARISON OF SOCIAL SYSTEMS

In applying comparison to behaviour, we have the good fortune to be already informed about the general outline of the natural system of affinity, and thus are in a much more favourable position than morphology was three hundred years ago. When we find, for instance, that the social organization underlying mating behaviour of the Cuttlefish is very similar to that of a fish, we do not believe for a moment that this is proof of real affinity between the two, because morphological study has already shown us that fish and Cuttlefish are not at all closely related. Certain morphological similarities such as 'fins' and eyes are convergences, and so are the similarities in mating pattern.

On the other hand, in comparing the mating patterns of closely related species, we may safely assume that their patterns are homologous. Thus when we see that the male Three-spined Stickleback in its zigzag dance first leads the female, and then attacks her; that the Ten-spined species first attacks her and then leads, and that the Sea Stickleback (*Spinachia vulgaris*) merely attacks her and does not lead her until she takes the initiative, we must assume that we have to do with three forms of basically the same behaviour pattern. When the mating patterns of closely related species are very different, we will be justified in trying to find the common root from which they have evolved.

Behaviour has rarely been studied systematically from this point of view. Yet social behaviour in particular offers a unique opportunity, because, owing to the need of reproductive isolation, there is usually a premium on diversity. This means that the social organization of related species has diverged very rapidly, and therefore closely related species—in which comparison is most easy because one can be certain of homologies—offer a wide spectrum of phenomena.

Just as in morphology, comparison can be practised at each level: at that of the community as a whole, at that of a minor system such as mating pattern, and at that of the single element of such a system, the single releaser. On each level just enough data are available to show the type of conclusions that can be drawn.

When the social systems of various types of bees are compared, we find that most species are solitary, and that the Honey Bee (and two related species) are exceptional in that they form highly complicated 'states' in which thousands of individuals co-operate. Since the social condition is clearly exceptional in the group, we conclude that bees were originally solitary. As related in Chapter VII there are groups, related to bees, which are social to a certain extent, and thus can be taken to have an intermediate position in this respect. By comparing solitary, intermediate, and highly social groups, as has been done to some extent in Chapter VII, it is found that the social organization has evolved from a family of mother and offspring, and that firstly an association of the mother with the brood, then a division of labour together with increasing complexity of co-operation developed.

Ants cannot add much to our theme, because there are no solitary ants. Neither are there solitary termites. Yet comparison of ants and termites reveals a striking example of far-reaching convergence. The social organization of termites has sprung from another source than that of ants or bees, for in termites

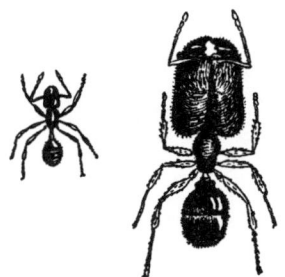 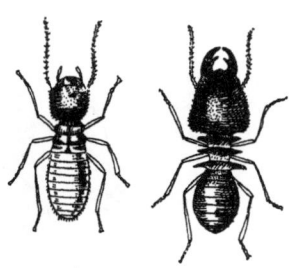

FIG. 64.—An ant worker (*left*) and soldier (*right*)

FIG. 65.—A termite worker (*left*) and soldier (*right*)

('social cockroaches') the males are represented in all castes; their states have evolved from a family formed by male and female with offspring. As is well known, the parallels between ants and termites go into many details; for instance 'soldiers' are found in both (Figs. 64 and 65).

Descending now from social organization as a whole to parts of it, we can again establish homologies and convergences. These are most striking in mating patterns. In visually well-equipped groups we often find sexual dimorphism, the males showing a conspicuous colour pattern, or performing special advertising ceremonies. Male Fiddler Crabs (Fig. 66),[16, 112] male Cuttlefish, male Fighting Fish (*Betta splendens*),[52] male lizards and male birds show this. They all use their conspicuous

colours to threaten other males, and their potentially ambivalent response—they either fight or court other members of
their species—is led into the purely sexual channel by special
reactions of the females. Some are predominantly aggressive,
such as *Betta*, or pigeons; others are predominantly sexual and
can only be stimulated to fight by special releasers in their
rivals. This is found in the Cuttlefish and in the Muscovy Duck.
In birds, the curious 'lek' system, in which many gorgeously
coloured males collect on a communal courting ground or lek
to which females come only for purposes of copulation, has

FIG. 66.—Male Fiddler Crab displaying (*after Pearse
from Verwey, 1930*)

been developed independently in the Ruff (*Philomachus pugnax*)
and in the Blackcock (*Lyrurus tetrix*).[46] In both species the males
do not join the females to help with the parental duties; there
is no personal bond whatsoever.

Within a genus, it is often possible to trace homologies. At
first glance, the pair formation behaviour of Herring Gull and
Black-headed Gull appears very different. The Herring Gull
pairs on the 'clubs' or social gathering ground, the Black-
headed Gull on 'pre-territories'. The unmated male Black-
headed Gull reacts very aggressively towards any stranger

whether male or female; the unmated male Herring Gull, while attacking other males, is not very aggressive towards females. The Black-headed Gull has an aerial display, the Herring Gull has not. The newly formed Black-headed Gull pair flies off to select a breeding territory, the newly formed Herring Gull pair walks away from the club and often selects a breeding territory not far from it. In some details there are considerable differences as well: the threat posture is different; the Herring Gull adopts the 'upright threat posture', the Black-headed Gull shows the 'forward display'. The appeasement postures are different: the Herring Gull adopts the 'submissive posture', the Black-headed Gull shows 'head flagging'.

Close analysis of the pair formation patterns in both species shows that they follow the same main plan: females approach males, appease them by showing the opposite of the threat posture; after mating-up the pair select a permanent territory together. The differences are related to two circumstances: (1) the Black-headed Gull, being the smaller species, resorts more to flying than the larger Herring Gull; this accounts for aerial displays not found in the Herring Gull, for the form of the threat posture (the upright threat posture is aimed at an opponent on the ground, the forward display at an opponent who may come either from the ground or from the air), and for the different type of departure to the permanent territory; (2) the threat posture of the Black-headed Gull is supported by the brown face, and this again accounts for the development of head flagging as an appeasing ceremony.

Our knowledge of these things is still very fragmentary, and entirely insufficient for a reconstruction of the historical processes through which the various types have evolved.

COMPARISON OF RELEASERS

More is known at a still lower level, that of the single signals. Here again, it is not difficult to detect homologies and convergences. The displacement preening of courting male duck, however different from one species to another, is certainly 'the same thing' throughout. The song of songbirds, again different from one species to another because of the demands of reproductive isolation, is homologous through the group, as is the instrument used, the syrinx. Examples of convergences are the frontal display of fish, which use the erected gill-covers, and of birds such as the Ruff, and the Domestic Cock, which use neck feathers to produce a gaudily coloured fan or collar.

Comparison of homologous releasers has led to remarkable conclusions about their origin and evolution.

So far, two sources of signal movements have been discovered. One is the intention movement. When duck or geese intend to fly up, their motivation is gradually built up. Very low motivation gives rise to incipient movements. The plumage is pressed against the body, and repeated bobbing of the head —the lowest intensity of the take-off—appears next. With growing motivation, the bobbing becomes more intensive, and other parts of the body may come into play as well: the wings are kept ready for action, the body may be slightly bent forward. Such low-intensity forms of the intended movement, act as releasers to the companions.

In other cases animals may make intention movements even when the motivation is rather high. The upright threat posture of a Herring Gull certainly signifies a rather strong tendency to attack. It does not easily develop into real attack because it is inhibited by a simultaneous tendency to flee or withdraw. Such inhibited intention movements act as signal movements in many other cases.

Displacement activities are the second source of releasers. Grass-pulling of Herring Gulls, displacement sand-digging of Three-spined Sticklebacks, showing-the-nest-entrance (displacement fanning) of the same species are examples; they all act as signals releasing certain responses in the opponent or in the sex partner.

It is difficult to see how such movements may have begun to be 'understood' by other individuals. This problem concerns the origin of the responsiveness of the reactor to the signal, not the origin of the signal movement itself. As regards intention movements, it is a problem of exactly the same order as the ultimate origin of an animal's responsiveness to any outside stimulus. It is just as mysterious why a Blackbird reacts to an

Earthworm, or to a Sparrow Hawk, as why it reacts to the intention movement of alarm shown by another Blackbird.

Why a Herring Gull 'understands' the aggressive nature of displacement collecting of nest-material (grass-pulling), and does not react to it by coming into the nest-building disposition itself, is a problem of another order. I believe there are two reasons why he should interpret it as aggressive behaviour. First, grass-pulling alternates with genuine aggressive behaviour. Second, as we have seen, the movement of grass-pulling is different from genuine collecting of nest-material: the gull pecks 'furiously' at the material, and pulls hard at it. These additions are parts of the fighting pattern; it treats the plants as if they were the opponent.

Once this responsiveness to a signal movement of another individual has been established, the further development of the signal function is an affair of both the actor and the reactor. In both, a new adaptive evolutionary process starts. Comparison has revealed to us several aspects of this process.[17] The displacement preening, a signal movement playing a part in the courtship of many male duck, is, in each species, not quite the same movement as the real preening. In some species real preening and displacement preening are so different that the displacement activity can hardly be recognized. Lorenz [56] has given accurate descriptions and illustrations, based on ciné-films, of displacement preening in many species (Fig. 67). The Mallard is a relatively primitive case. The male just brings its bill behind the wing, much as in an ordinary preening movement, though the displacement preening is more stereotyped. The Mandarin Drake has a very specialized movement: it deftly touches the vane of one of the secondaries. This secondary is not just a dark green plume as the other secondaries are. Its outer vane has developed into a huge flag-like structure, and its colour is bright orange instead of green. The Garganey Drake has a different movement again. It does not touch the inner side of the wing but the outside, exactly at the spot where the wing coverts are a bright greyish-blue. In both the Garganey and the Mandarin, therefore, a conspicuous structure has developed and the movement now draws attention to this structure. This whole

development results in making the movement more conspicuous and stereotyped; it has become a 'rite'. At the same time,
the movements of different species here diverged; they have
become more specific. The evolutionary process which 'gets

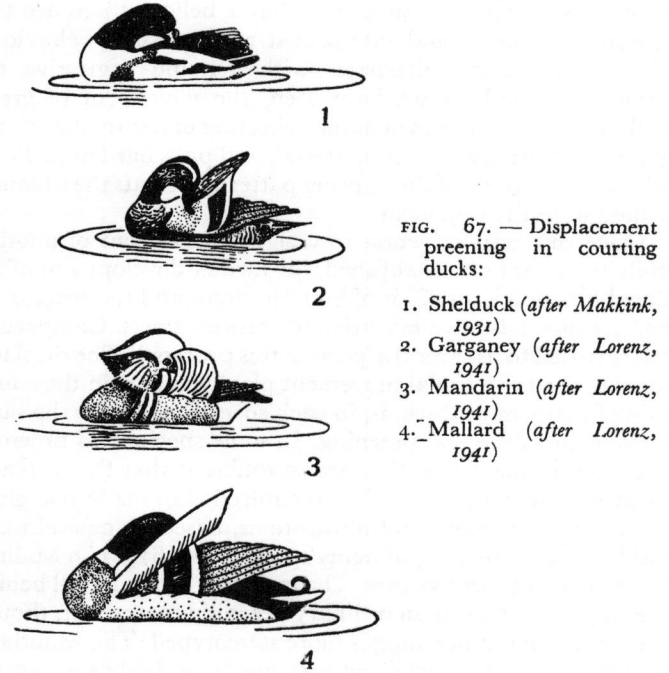

FIG.   67. — Displacement
  preening   in   courting
  ducks:

1. Shelduck (*after Makkink,
     1931*)
2. Garganey (*after Lorenz,
     1941*)
3. Mandarin (*after Lorenz,
     1941*)
4. Mallard (*after Lorenz,
     1941*)

hold' of signals and makes them more conspicuous and more
specific is called ritualization.

All evidence so far known points to the conclusion that signal
movements have originally been movements without signal
function; they were, in a way, 'by-products' of nervous organization. After they acquired signal function, a new type of adaptive evolution, ritualization, started, which led to a change
of the movement and a coincident change in morphological
structures.

Ritualization is adaptive in two respects. Ritualized releasers are always characterized by conspicuousness and simplicity. This is an adaptation to the limitations of the responsiveness of innate behaviour. Each innate reaction depends for its release on specific stimuli; a study of the stimuli required has shown in a number of cases that these stimuli are always relatively simple and conspicuous. Ritualization tends to make the releaser specialize into just showing such 'sign stimuli'; releasers are, in a way, 'materialized sign stimuli'. Second, ritualization tends to make the releaser different from any other releaser, whether of the same species or of other species. It thus facilitates social co-operation within the species, and it reduces the chance of response to other species.

In both intention movements and displacement activities ritualization seems to have followed the same lines. In both, the emphasis has sometimes been on movement, sometimes on structure. One of the most common changes in a movement is 'schematizing' which implies exaggeration of some parts of the movement and obliteration of other parts. This has occurred, for instance, in some of the courtship movements of ducks. For example, 'shortening-up' meant originally a lifting of the head and of the tail. In the Garganey the back movement of the head has been emphasized, whereas the upward movement of the tail has entirely disappeared. In the Chilean Teal, the head movement has developed in another way; the main emphasis is on the movement of the breast, and the tail is not involved either. In the Pintail, both head and tail take part in the movement; the tail movement is here supported by the conspicuously coloured triangle at the tail base, and by the elongated tail itself.

Physiologically, these and some other aspects of ritualization can be understood as quantitative changes in thresholds of the various component parts of the movement. A fuller treatment of these problems would carry us beyond the scope of this book. I should like to point out, however, that the study of the evolution of releasers has a very important bearing on the problem of the origin and evolution of 'new' behaviour elements in general, for the ritualization of intention or displacement movements does lead to the evolution of new movements. Again, this cannot be worked out here.

CONCLUSION

The work so far done on the origin and evolution of releasers, particularly those playing a part in courtship and threat, makes it increasingly clear that they were originally accidental by-products, outlets of nervous excitation either in the form of intention movements or of displacement activities. In most cases, these outlets appear when the normal outlet is hampered by the simultaneous activation of antagonistic drives. This holds for all the 'inhibited intention movements', and for all displacement activities used as threat, and possibly for many displacement activities used in courtship. This seems to give us the clue as to why 'display' is so widespread exactly in courtship and in threat. In courtship, while the sex drive may be the main component of the motivation, aggressiveness and a tendency to escape play a part as well. In threat,[103] aggressiveness and the tendency to escape are in conflict with each other. We have seen that both aggressiveness and sexual behaviour are necessary for the maintenance of the species. Neither can be omitted. Because innate behaviour responds to simple sign stimuli, and since the female, as a member of the same species, cannot help providing sign stimuli releasing attack besides giving stimuli releasing a sexual response, a male is always stimulated both aggressively and sexually by an approaching female. If the male's aggressiveness were less, he would be able to show purely sexual responses to the female, but then he would not be successful in fighting off other males. If his sexual drive were stronger, this would override not only his aggressiveness to females, but also all his other drives, such as escape from predators. If his escape drive were weaker, this would make him a very successful fighter but it would endanger his escape from predators. There is, in each animal, a fair balance between the various drives. Threat and courtship are the inevitable consequences of this balance; by ritualization they have been put to as good a use as was possible under the circumstances.

CHAPTER IX

SOME HINTS FOR RESEARCH IN ANIMAL SOCIOLOGY

A GLANCE at the names in the bibliography shows that animal sociology owes much to the work of 'amateurs'. Selous, Howard, Portielje, to mention only a few workers whose contributions have done so much to develop this field, were and are no professional zoologists. As a matter of fact, official zoology has long left animal sociology alone, and the early work has all been done either by amateurs or by zoologists who had no training in this type of work at all. Both Heinroth and Huxley were, as animal sociologists, autodydacts when they wrote their pioneer contributions. Thanks to their work, and the subsequent work of Lorenz and his co-workers, interest is now rapidly growing among zoologists. This leads to a quicker development, and this again results in the introduction of new concepts and terms, and in a rapid growth of the literature. This certainly is encouraging, but it has the disadvantage that research becomes increasingly the monopoly of professional specialists. Many amateurs feel that they can no longer keep pace with it, let alone produce new and original contributions. I don't think such pessimism is justified. It is not only possible, it is also very desirable that non-professionals go on to contribute, for lack of specialized training has advantages as well as disadvantages. Of course training gives knowledge and discipline of thought, but it often tends to smother originality of outlook. The amateur may approach the subject with a certain freshness of mind which may have a profound influence. This final chapter will give some hints to those who feel inclined to embark upon some work on their own.

It is obvious that the best contributions have come from people who have given years of their life to careful, patient observation of one species. Comparison of several species, whether closely related or not, has also been very important, but this can only be done after a profound knowledge of one species has been acquired.

The need for a broad, observational approach cannot be stressed too much. The natural tendency of many people, particularly of young beginners, is to concentrate on an isolated problem and to try to penetrate into it. This laudable inclination must be kept in check or else it leads to an accumulation of partial, disconnected results, to a collection of sociological oddities. A broad, descriptive reconnaissance of the whole system of phenomena is necessary in order to see each individual problem in its perspective; it is the only safeguard for a balanced approach in which analytical and synthetical thinking can cooperate. This, of course, is true not only of sociology, it is true of each science, but in ethology and sociology it is perhaps forgotten more often than in other sciences.

Since this broad observational approach is, in my opinion, of such extreme importance, I will elaborate it a little. I was once visited by a keen student from abroad who wanted to receive training in sociological work. He arrived with one very special problem in mind: he wanted to be trained in the technique of the experimental study of releasers. I tried in vain to convince him that he could better begin with a broad reconnaissance of a species; then let him have his way, and he started to count the number of bites aimed by a territory-owning male Three-spined Stickleback at a red model as compared with a silvery model. His results seemed to be at variance with our previous work: the red models received only slightly more bites than the silver models. On doing the tests again it was found that the fish showed several signs of hostility other than actual bites (such as raising the dorsal spines, and making incipient attacks) and that these were released by red models much more often than by the silver models. Having skipped the observational study of aggressive behaviour he had been unable to recognize and interpret these hostile movements. He then returned to just watching, and when, after some days, he resumed his tests, he got very clear-cut results.

Displacement activities provide another example. Without an observational knowledge of both the pattern of the drive that uses the displacement activity as an outlet and the pattern from which it is 'borrowed' it is impossible to understand a displacement activity and to see the nature of its connexions with the two drives.

The head-flagging of Black-headed Gulls, which, as I mentioned, is an appeasement-gesture, becomes intelligible only when its counterpart, the forward threat posture, is known as well. Unless fighting behaviour is studied as well as courtship, an observer must fail to understand the head flagging. Also, ignoring the threat behaviour will prevent one realizing the important fact that courtship is always mingled with aggressive tendencies.

The revival of song, by which a female Phalarope entices the male to follow her to the nest when she is going to lay an egg, cannot be understood unless it is known that the same call attracts unmated males to territorial females early in the season, and even then the egg-ceremony does not make sense unless one knows that the male incubates alone and has to be shown where the female is going to deposit the eggs. This again becomes intelligible when it is known that the rôles played by the sexes are reversed as well as the sexual dimorphism of the plumage.

These are only a few examples. Though it requires some, and sometimes considerable, self-restraint to stick to the broad observational reconnaissance before one tackles detailed problems, and although the reconnaissance may lead to no clear-cut 'results' for a long time, perseverance is ultimately rewarded, and gradually things will begin to 'make sense', and problems are seen everywhere, and in their proper relations to other problems.

Repetition of the observations is also of great importance. Social behaviour involves so many things happening at the same time, that it is impossible to see everything. Your attention must be with the actor and with the reactor, and with other individuals in the vicinity as well. Even the movements of one individual cannot be grasped if seen only once, let alone all that happens simultaneously. Only by watching, writing down, drawing, realizing how much you are not certain about, watching again, and thus completing your description step by step, can you attain a reasonable accuracy and completeness. I am not exaggerating when I say that I have watched the courtship of the Three-spined Stickleback hundreds of times, and still I am seeing new details, some of which contribute to a better understanding of basic problems. Films are a great help for this. A good ciné-record of one particular event can have the same value as many hours or even days spent in watching.

A great deal of observational work can be done in the field with wild animals. The advantage is that the animals are then in their proper environment—which is often very difficult to imitate in captivity—they are in perfect health, and they require no care; Nature just keeps them for us. The setback of shyness can easily be overcome by the use of hides. Field work has given excellent results mainly with birds and with insects. Many of the data on birds mentioned in this book have been collected in the field: Makkink's work on the Avocet, that of Kortlandt on the Cormorant, Laven's observations on the Ringed Plover, Lack's work on the Robin, that of my own on the Herring Gull—these and many others are entirely based on field work. The equipment for such studies is simple. Binoculars are almost indispensable. For continuous watching it is very important to mount them on a tripod with pan-and-tilt-head. Your hands will inevitably begin to tremble after you have been watching for an hour or so, but even before that your glasses will move slightly with each beat of your pulse; it is amazing how much more you can see when this is eliminated by mounting your glasses. If you have no tripod, just put your glasses on some stones or on a gate or a tree, and put a stone on top of them.

A second accessory of the field watcher is some way of marking individuals. Without marking them you can of course often recognize individuals by peculiarities of plumage, by a damaged leg, by abnormal size, &c., but such animals are always recognizable just because they have some slight abnormality, and the possibility exists that exactly these animals may show, or release in others, abnormal behaviour. Students of migration have long used numbered aluminium rings. The figures on them are usually so small however, that they cannot be read from a distance. For very large birds, such as Storks, rings are used with large numbers that can be read in the field, but for smaller birds coloured rings are the solution. By combining five or six different colours one can mark a great number of birds individually. There is nothing against giving a bird two or even three rings on each leg, dependent on the species. Some of my marked Herring Gulls jingled merrily every time they flew off, but they seemed not to mind at all and lived for years.

For some observations, and for the taking of photographs or ciné-films, it is necessary to use hides. I use collapsible canvas tents of 4 cubic feet, with a metal framework of units 2 feet long. They take a few minutes to put up, can easily be transported, and when properly fixed withstand strong wind. It is advisable to camouflage observation windows with plants at the outside. These break up the dark hole by the irregular leaves, which are themselves in broad daylight. The observer can then make any movements inside without being noticed. In such hides it is very important not to open windows simultaneously at opposite sides, since the birds may then see your silhouette move across the window behind you.

However, for much observational work it is decidedly better to sit in the open so that you can have a wide look round, for it is often just as important to see to what events in the vicinity your birds are reacting as to see what they do themselves. You have to keep so far away from your birds that they do not mind your presence. This distance can often be surprisingly small once the birds are used to you, which happens soon when you keep quiet. They then do not give you more attention than they do to a cow, a very desirable status for the bird watcher.

For bird work one has to get up early. Most birds show a maximum of activity, particularly of reproductive behaviour, in the hours round sunrise. A second, but lower maximum falls in the evening. It is best to arrive about an hour before sunrise, and stay till three or four hours after sunrise, when the activity wanes. Once you get used to being in the field early, you will like it much better than getting out later when the sun has already climbed high, the dew has evaporated, and the scenery has become dry, colourless, and dull. Further, the more promptly one reacts to the alarm clock, the easier it is.

Insects can also be studied in the field. In many respects they are even better objects than birds. They are far less shy, and their maximum of activity is not in the early morning hours, and therefore it is less strenuous to make continuous, day-to-day studies. Energetic people can begin their days with birds, and shift to insects at about 9 a.m.

The great French observer Fabre has shown how much of interest can be revealed by simple watching. His work, however valuable in his time, is not of sufficient precision for our present purposes. A modern study suited to demonstrate the type of results one can get is that by Baerends on the behaviour of the Digger Wasp *Ammophila adriaansei*. In this species he discovered very complicated relationships between the females and their offspring. Each larva lives solitarily in a burrow and is provided with paralysed caterpillars by the mother. Baerends not only observed the normal course of events in detail, marking both nests and individuals, but he carried out extensive experiments. For instance, he discovered that each female can take care of two or even three nests simultaneously, each at a different stage of development, and that she knows exactly when any of the larvae needs a new supply of food. He substituted plaster burrows for the real burrows so that he could open a nest whenever he wanted, and change its contents. In this way he could prove that the behaviour of a female was influenced by the contents of the burrow, such as the amount of food present, and the age of the larva.

Insects offer an almost unlimited field of research. The work of Baerends shows what fascinating objects the Digger Wasps are. With butterflies we have made no more than a promising beginning; the Grayling work shows what we can expect. Dragonflies are another interesting group; the beautiful species *Calopteryx virgo* for instance, has developed a type of mating behaviour very similar to that found in many birds and fish; the males defend a territory against other males; also, they have a specialized courtship, entirely based on visual stimulation. Grasshoppers and Locusts have developed social relationships of quite another type, as the work of Jacobs [35a] and of Duym and Van Oyen [20] has shown. In other groups such as mammals, the work done so far (Schenkel,[78a] Hediger,[29] Carpenter,[12a] Eibl von Eibesfeldt [20a, 20b]) lizards (Noble,[66] Kramer,[42] Kitzler,[38]) and spiders (Crane,[16a, 16b]) though not so extensive yet as the bird work, allows interesting comparisons, and these groups certainly deserve more attention than they have received so far.

The zoological gardens are another medium for sociological research. The animals can here be observed at close quarters, and often the more or less abnormal environment gives rise to deviations from the normal course which are of great value to our understanding of the natural events. Also, comparisons with exotic species can be made which often are beyond the range of the field observer. Heinroth, who worked in Berlin, and Portielje in Amsterdam are the pioneers of this type of work, and the respectable series of their publications demonstrates the value of zoos for sociology. The importance of zoos for behaviour studies is now becoming generally recognized; in Switzerland for instance both the Bâle and the Berne zoos are directed by behaviour specialists.

A special, and for our purpose very valuable, type of zoo is the aquarium. Valuable because it is by far the cheapest way of keeping animals in approximately natural surroundings and is within reach of almost everybody. In fact, once you have an aquarium of modest dimensions (e.g. 18 × 12 × 12 inches), you can without any cost observe everything said in this book (and much more) about the Three-spined Stickleback or the Ten-spined Stickleback. You will have to spend some hours catching your fish in early spring, and you will have to dig up a worm every day, but that is all. Many of our indigenous fish have not yet been studied, and the various newts also are worth a more detailed study. From a freshwater tank to a sea-water aquarium it is but a small step; also, it is possible with very little expense to instal a tropical tank, and study any of the numerous tropical species that have been imported. The field is practically unlimited. Many groups of fish have developed highly specialized systems of visual releasers, and their capacity

II

for changing colour rapidly makes them still more fascinating than many birds.

Lorenz has developed a special type of zoo. He has raised and kept a number of animals in a kind of semi-captivity. They are allowed, within certain (very wide) limits, to move about freely, and by raising them personally he has tied them socially to him. Many of these animals treat him as a member of their own species: they court or fight him, or try to make him join them when they move about. This opens unique opportunities for study, which Lorenz has utilized to the utmost, literally living with his animals from day to day. For the information of anybody who might feel inclined to start similar studies, I should add that they cannot be undertaken without the consent of the housewife in charge.

The observational work has to be followed up by experimental study. This can often be done in the field. The change from observation to experiment has to be a gradual one. The investigation of causal relationships has to begin with the utilization of 'natural experiments'. The conditions under which things occur in nature vary to such a degree that comparison of the circumstances in which a certain thing happens often has the value of an experiment, which has only to be refined in the crucial tests. For instance, Heinroth's observation of a Swan attacking its mate when the latter's head happened to be submerged indicates that the characters which allow individual recognition must be located in the head, and this provides one with the basis for more exact experiments. The fact that a male Stickleback guides a female to the nest, but chases her immediately after she has spawned, suggests that the swollen abdomen which she had before the spawning may have something to do with the release of his courtship. When I observed repeatedly in the field that female Phalaropes courted passing Ringed Plovers, Lapland Longspurs, and Purple Sandpipers, but never reacted to Snow Buntings—the only one of these species which has a striking white patch on the wing—this suggested that the courting was released by any bird of approximately the same dull colour pattern as the Phalarope. A field observer encounters many such natural experiments in the course of one day, and a systematic watch for them will supply him with an extensive programme for experimentation. Although in experiments with dummies morphological characters, such as colour and shape, are easily imitated and varied, movement is very difficult to imitate, and evidence on the influence of type of movement has often to be based entirely on long series of 'natural experiments'.

Captive animals naturally invite more experimentation than free living animals, because they cannot get away from the experiment even if they want. But this implies a certain danger, for it tempts the observer to overdo his experiments. Experimenting is, in more than one way, a delicate business. First of all, the animal must be in the appropriate 'mood'. It is little use to offer a model of the bill of an adult Herring Gull to a chick that has just been alarmed by the warning call of the adults, or to one that has just been fed. The most obvious disturbing factor is the escape response. It is only too easy to evoke escape tendencies. In the most clear-cut cases these are easily inferred, for overt escape behaviour can rarely be mistaken. But even a weak activation of the escape drive inhibits other behaviour, and it requires sharp observation and considerable experience with a given species to detect slight signs of inhibition through fear. This is not astonishing when we realize how many people fail to recognize even rather obvious expressions in their fellow-men, and also how much more difficult it is to recognize such expression in species other than our own.

Each experiment has to be repeated a number of times to eliminate the influence of variables beyond the control of the experimenter. It is always tempting to use one individual for more than one test, rather than to take a new individual for each test. Here, however, one must make sure that the animal does not change during the test series. One common cause of change is exhaustion of the drive involved, which causes a progressive decrease of responsiveness. This often happens when the individual tests are done with too short intervals. Another cause is learning. Young Herring Gulls which were presented again and again with head models to which they reacted without ever getting food, became negatively conditioned to them and gave fewer and fewer responses. Geese which were shown cardboard models of a bird of prey, sailing overhead, became positively conditioned to the experimental set-up and began to call the alarm every time the experimenter, as a preparation for a test, climbed a tree to fasten the model.

This leads us to the necessity of controls. Each experiment is a comparison of the effect of two situations, differing in the one aspect the influence of which one wants to study. When, for instance, one wants to know which stimuli from the eggs release incubation and which don't, it is not sufficient just to show that a bird will accept an abnormal egg. The reaction to the abnormal egg must be compared with the reactions to normal eggs; if there is a difference in the intensity or kind of the reactions, this means that the difference contains an element which influences the bird's reaction. A test with an abnormal egg without any control test is sufficient to conclude that the abnormal egg contains some stimuli which influence incubation, but it does not show that the abnormal egg provides all the stimuli. This might seem a truism, but it has to be emphasized since several studies published in scientific journals of standing suffer from this defect.

These are the main pitfalls. It is not possible to give more than general directions. The sources of error mentioned may work out in an endless variety of ways, and it is often a matter of 'intuition' to recognize and appreciate them. The trick is, to insert experiments now and then in the normal life of the animal so that this normal life is in no way interrupted; however exciting the result of a test may be for us, it must be a matter of daily routine to the animal. A man who lacks the feeling for this kind of work will inevitably commit offences just as some people cannot help kicking and damaging delicate furniture in a room without even noticing it.

Publishing results of work is an essential part of the investigation. Good contributions are welcome in most zoological journals. The international journal *Behaviour* is perhaps the most appropriate channel. Work on birds is often published in ornithological journals, of which the *Ibis* is the obvious one for the British worker. Simplicity and straightforwardness of language are essential; not only to the reader but also to the author; often the writing down of a study is a considerable help in organizing one's thoughts and seeing the problems clearly. Illustrations are a most important element of publications of this kind. It is impossible to describe complicated behaviour types in sufficient detail and yet in such a way that the reader can visualize it. One mediocre drawing or photograph is often more useful than two pages of accurate but necessarily dull description. The observer should make sketches while in the field and keep checking and improving them. Ciné-films are of great help here, in fact, they are almost essential for accurate work; they can be used as a basis of drawings. For reasons of economy drawings should be made fit for line blocks, for most scientific journals fight a continuous struggle against bankruptcy.

In most cases publication cannot be undertaken without a certain amount of reading. It should be stressed that in order to get really informed about the recent st̄te of knowledge it is not sufficient to confine reading to the English language. The serious student of sociology cannot do without the continental literature, which in our field is mainly written in German. The work of Heinroth, Lorenz, Koehler and of their followers and pupils is essential and has not penetrated in full into the English literature. Much of it is to be found in the *Journal für Ornithologie* and in the *Zeitschrift für Tierpsychologie*.

On the other hand, I must point out that extensive reading, however necessary, can never replace first-hand knowledge based on one's own watching. The animals themselves are always more important than the books that have been written about them.

BIBLIOGRAPHY

[1] ALLEE, W. C., 1931: *Animal Aggregations*. Chicago.

[2] ALLEE, W. C., 1938: *The Social Life of Animals*. London-Toronto.

[3] BAERENDS, G. P., 1941: 'Fortpflanzungsverhalten und Orientierung der Grabwespe Ammophila campestris. Jur.' *Tijdschr. Entomol.*, **84**, 68–275.

[4] BAERENDS, G. P., 1950: 'Specializations in organs and movements with a releasing function'. *Symposia of the S.E.B.*, **4**, 337–60.

[5] BAERENDS, G. P., and BAERENDS, J. M., 1948: 'An introduction to the study of the ethology of Cichlid Fishes'. *Behaviour, Suppl.*, **1**, 1–242.

[6] BATES, H. W., 1862: 'Contributions to an insect fauna of the Amazon Valley'. *Trans. Linn. Soc.*, London, **23**, 495–566.

[7] BEACH, F. A., 1948: *Hormones and Behavior*. New York.

[8] BOESEMAN, M., VAN DER DRIFT, J., VAN ROON, J. M., TINBERGEN, N., and TER PELKWIJK, J., 1938: 'De bittervoorns en hun mossels'. *De Lev. Nat.*, **43**, 129–236.

[9] BULLOUGH, W. S., 1951: *Vertebrate Sexual Cycles*. London.

[10] BURGER, J. W., 1949: 'A review of experimental investigations of seasonal reproduction in birds'. *Wilson Bulletin*, **61**, 201–30.

[11] BUXTON, J., 1950: *The Redstart*. London.

[12] CINAT-TOMSON, H., 1926: 'Die geschlechtliche Zuchtwahl beim Wellensittich (*Melopsittacus undulatus Shaw*)'. *Biol. Zbl.*, **46**, 543–52.

[12a] CARPENTER, C. R., 1934: 'A field study of the behavior and social relations of Howling Monkeys'. *Comp. Psychol. Mon.*, **10**, 1–168.

[13] COTT, H., 1940: *Adaptive Coloration in Animals*. London.

[14] CRAIG, W., 1911: 'Oviposition induced by the male in pigeons'. *Jour. Morphol.*, **22**, 299–305.

[15] CRAIG, W., 1913: 'The stimulation and the inhibition of ovulation in birds and mammals'. *Jour. anim. Behav.*, **3**, 215–21.

[16] CRANE, J., 1941: 'Crabs of the genus Uca from the West Coast of Central America'. *Zoologica*, N.Y., **26**, 145–208.

[16a] CRANE, J., 1949: 'Comparative biology of salticid spiders at Rancho Grande, Venezuela. IV. An analysis of display'. *Zoologica N.Y.*, **34**, 159–214.

[16b] CRANE, J., 1949: 'Comparative biology of salticid spiders at Rancho Grande, Venezuela. III. Systematics and behavior in representative new species'. *Zoologica N.Y.*, **34**, 31–52.

[17] DAANJE, A., 1950: 'On locomotory movements in birds and the intention movements derived from them'. *Behaviour*, **3**, 48–98.

[18] DARLING, F. F., 1938: *Bird Flocks and the Breeding Cycle*. Cambridge.

[19] DICE, L. R., 1947: 'Effectiveness of selection by owls of deer-mice (*Peromyscus maniculatus*) which contrast in color with their background'. *Contr. Lab. Vertebr. Biol.*, Ann Arbor, **34**, 1–20.

[20] DUYM, M., and VAN OYEN, G. M., 1948: 'Het sjirpen van de Zadelsprinkhaan'. *De Levende Natuur*, **51**, 81–7.

[20a] EIBL-EIBESFELDT, I., 1950: 'Ueber die Jugendentwicklung des Verhaltens eines männlichen Dachses (*Meles meles* L.) unter besonderer Berücksichtigung des Spieles'. *Zs.f. Tierpsychol.*,**7**, 327–55.

[20b] EIBL-EIBESFELDT, I., 1951: 'Beobachtungen zur Fortpflanzungsbiologie und Jugendentwicklung des Eichhörnchens (*Sciurus vulgaris* L.)'. *Zs. f. Tierpsychol.*, **8**, 370–400.

[21] FABRICIUS, E., 1951: 'Zur Ethologie junger Anitiden'. *Acta Zoologica Fennica*, **68**, 1–177.

[22] FRISCH, K. VON, 1914: 'Der Farbensinn und Formensinn der Biene'. *Zool. Jahrb. Allg. Zool. Physiol.*, **35**, 1–188.

[23] FRISCH, K. VON, 1938: 'Versuche zur Psychologie des Fisch-Schwarmes'. *Naturwiss.*, **26**, 601–7.

[24] FRISCH, K. VON, 1950: *Bees, their Vision, Chemical Senses, and Language.* Ithaca, N.Y.

[25] GOETHE, FR., 1937: 'Beobachtungen und Untersuchungen zur Biologie der Silbermöwe (*Larus a. argentatus*) auf der Vogelinsel Memmertsand'. *Jour f. Ornithol.*, **85**, 1–119.

[26] GOETSCH, W., 1940: *Vergleichende Biologie der Insektenstaaten.* Leipzig.

[27] GÖZ, H., 1941: 'Über den Art- und Individualgeruch bei Fischen'. *Zs. vergl. Physiol.*, **29**, 1–45.

[28] GRASSÉ, P. P., and NOIROT, CH.: 'La sociotomie: migration et fragmentation chez les Anoplotermes et les Trinervitermes'. *Behaviour*, **3**, 146–66.

[29] HEDIGER, H., 1949: 'Säugetier-Territorien und ihre Markierung'. *Bijdr. tot de Dierk.*, **28**, 172–84.

[30] HEINROTH, O., 1911: 'Beiträge zur Biologie, namentlich Ethologie und Psychologie der Anatiden'. *Verh. V. Intern. Ornithol. Kongr.*, Berlin, 589–702.

[31] HEINROTH, O., and HEINROTH, M., 1928: *Die Vögel Mitteleuropas.* Berlin.

[32] HINDE, R., 1952: 'Aggressive behaviour in the Great Tit'. *Behaviour, Suppl. 2*, 1–201.

[33] HOWARD, H. E., 1920: *Territory in Bird Life.* London.

[34] HUXLEY, J. S., 1934: 'Threat and warning coloration in birds'. *Proc. 8th Internat. Ornithol. Congr.*, Oxford, 430–55.

[35] ILSE, D., 1929: 'Über den Farbensinn der Tagfalter'. *Zs. vergl. Physiol.*, **8**, 658–92.

[35a] JACOBS, W., 1948: 'Vergleichende Verhaltensforschung bei Feldheuschrecken'. *Verh. d. deutschen Zool. Gesellsch.*, 1948, 257–62.

[36] JONES, F. M., 1932: 'Insect coloration and the relative acceptability of insects to birds'. *Trans. Entomol. Soc.*, London. **80**, 345–85.

[37] KATZ, D., and RÉVÉSZ, G., 1909: 'Experimentell-psychologische Untersuchungen mit Hühnern'. *Zs. Psychol.*, **50**, 51–9.

[38] KITZLER, G., 1941: 'Die Paarungsbiologie einiger Eidechsenarten'. *Zs. f. Tierpsychol.*, **4**, 353–402.

[39] KNOLL, FR., 1926: *Insekten und Blumen*. Wien.

[40] KNOLL, FR., 1925: 'Lichtsinn und Blütenbesuch des Falters von Deilephila livornica'. *Zs. vergl. Physiol.*, **2**, 329–80.

[41] KORRINGA, P., 1947: 'Relations between the moon and periodicity in the breeding of marine animals'. *Ecol. Monogr.*, **17**, 349–81.

[42] KRAMER, G., 1937: 'Beobachtungen über Paarungsbiologie und soziales Verhalten von Mauereidechsen'. *Zs. Morphol. Oekol. Tiere*, **32**, 752–84.

[43] KUGLER, H., 1930: 'Blütenökologische Untersuchungen mit Hummeln. 1'. *Planta*, **10**, 229–51.

[44] LACK, D., 1932: 'Some Breeding habits of the European Nightjar'. *The Ibis*, Ser. 13, **2**, 266–84.

[45] LACK, D., 1933: 'Habitat selection in birds'. *Jour. anim. Ecol.*, **2**, 239–62.

[46] LACK, D., 1939: 'The display of the Blackcock'. *Brit. Birds*, **32**, 290–303.

[47] LACK, D., 1943: *The Life of the Robin*. London.

[48] LACK, D., 1947: *Darwin's Finches*. Cambridge.

[49] LAVEN, H., 1940: 'Beiträge zur Biologie des Sandregenpfeifers (*Charadrius hiaticula L.*)'. *Jour. f. Ornithol.*, **88**, 183–288.

[50] LEINER, M., 1929: 'Oekologische Untersuchungen an *Gasterosteus aculeatus L.*' *Zs. Morphol. Oekol. Tiere*, **14**, 360–400.

[51] LEINER, M., 1930: 'Fortsetzung der oekologischen Studien an *Gasterosteus aculeatus L.*' *Zs. Morphol. Oekol. Tiere*, **16**, 499–541.

[52] LISSMANN, H. W., 1932: 'Die Umwelt des Kampffisches (*Betta splendens Regan*)'. *Zs. vergl. Physiol.*, **18**, 65–112.

[53] LORENZ, K., 1927: 'Beobachtungen an Dohlen'. *Jour. f. Ornithol.*, **75**, 511–19.

[54] LORENZ, K., 1931: 'Beiträge zur Ethologie sozialer Corviden'. *Jour. f. Ornithol.*, **79**, 67–120.

[55] LORENZ, K., 1935: 'Der Kumpan in der Umwelt des Vogels'. *Jour. f. Ornithol.*, **83**, 137–213 and 289–413.

[56] LORENZ, K., 1941: 'Vergleichende Bewegungsstudien an Anatinen'. *Jour. f. Ornithol.*, **89** (Festschrift Heinroth), 194–294.

[57] LORENZ, K., 1952: *King Solomon's Ring*. London.

[58] MCDOUGALL, W., 1933: *An Outline of Psychology*. 6th ed. London.

[59] MAKKINK, G. F., 1931: 'Die Kopulation der Brandente (*Tadorna tadorna L.*)'. *Ardea*, **20**, 18–22.

[60] MAKKINK, G. F., 1936: 'An attempt at an ethogram of the European Avocet (*Recurvirostra avosetta L.*) with ethological and psychological remarks'. *Ardea*, **25**, 1–60.

[61] MARQUENIE, J. G. M., 1950: 'De balts van de Kleine Watersalamander'. *De Lev. Nat.*, **53,** 147–55.

[62] MATTHES, E., 1948: 'Amicta febretta. Ein Beitrag zur Morphologie und Biologie der Psychiden'. *Mémor. e estudos do Mus. Zool., Coimbra,* **184,** 1–80.

[63] MEISENHEIMER, J., 1921: *Geschlecht und Geschlechter im Tierreich*. Jena.

[64] MOSEBACH-PUKOWSKI, E., 1937: 'Über die Raupengesellschaften von *Vanessa io* und *Vanessa urticae*'. *Zs. Morphol. Oekol. Tiere,* **33,** 358–80.

[65] MOSTLER, G., 1935: 'Beobachtungen zur Frage der Wespenmimikry'. *Zs. Morphol. Oekol. Tiere,* **29,** 381–455.

[66] NOBLE, G. K., 1934: 'Experimenting with the courtship of lizards'. *Nat. Hist.,* **34,** 1–15.

[67] NOBLE, G. K., 1936: 'Courtship and sexual selection of the Flicker (*Colaptes auratus luteus*)'. *The Auk,* **53,** 269–82.

[68] NOBLE, G. K., and BRADLEY, H. T., 1933: 'The mating behaviour of lizards'. *Ann. N.Y. Acad. Sci.,* **35,** 25–100.

[69] NOBLE, G. K., and CURTIS, B., 1939: 'The social behavior of the Jewel Fish, *Hemichromus bimaculatus Gill*'. *Bull. Am. Mus. Nat. Hist.,* **76,** 1–46.

[70] PELKWIJK, J. J. TER, and TINBERGEN, N., 1937: 'Eine reizbiologische Analyse einiger Verhaltensweisen von *Gasterosteus aculeatus L.*' *Zs. f. Tierpsychol.,* **1,** 193–204.

[71] PORTIELJE, A. F. J., 1928: 'Zur Ethologie bzw. Psychologie der Silbermöwe (*Larus a. argentatus Pontopp.*)'. *Ardea,* **17,** 112–49.

[72] POULTON, E. B., 1890: *The Colours of Animals.* London.

[73] RIDDLE, O., 1941: 'Endocrine aspects of the physiology of reproduction'. *Ann. Rev. Physiol.,* **3,** 573–616.

[74] ROBERTS, BR., 1940: 'The breeding behaviour of penguins'. *Brit. Graham Land Exped., 1934–1937. Scientif. Reports* **1,** 195–254.

[75] ROESCH, G. A., 1930: 'Untersuchungen über die Arbeitsteilung im Bienenstaat'. 2. Teil. *Zs. vergl. Physiol.,* **12,** 1–71.

[76] ROWAN, W., 1938: 'Light and seasonal reproduction in animals'. *Biol. Rev.,* **13,** 374–402.

[77] BLEST, A. D., and DE RUITER, L.: Unpublished work.

[78] RUSSELL, E. S., 1945: *The Directiveness of Organic Activities.* Cambridge.

[78a] SCHENKEL, R., 1947: 'Ausdrucks-Studien an Wölfen'. *Behaviour,* **1,** 81–130.

[79] SCHREMMER, FR., 1941: 'Sinnesphysiologie und Blumenbesuch des Falters von *Plusia gamma L.*'. *Zool. Jahrb. Syst.,* **74,** 375–435.

[80] SCHUYL, G., TINBERGEN, L., and TINBERGEN, N., 1936: 'Ethologische Beobachtungen am Baumfalken, *Falco s. subbuteo L.*'. *Jour. f. Ornithol.,* **84,** 387–434.

[81] SCOTT, P., 1951: *Third Annual Report, 1949–1950, of the Severn Wildfowl Trust.* London.

[82] SEITZ, A., 1941: 'Die Paarbildung bei einigen Cichliden II'. *Zs. f. Tierpsychol.*, **5**, 74-101.

[83] SEVENSTER, P., 1949: 'Modderbaarsjes'. *De Lev. Nat.*, **52**, 161-68, 184-90.

[84] SPIETH, H. T., 1949: 'Sexual behavior and isolation in Drosophila II. The interspecific mating behavior of species of the willistoni-group'. *Evolution*, **3**, 67-82.

[85] SUMNER, F. B., 1934: 'Does "protective coloration" protect?' *Proc. Acad. Sci. Washington*, **20**, 559-564.

[86] SUMNER, F. B., 1935: 'Evidence for the protective value of change-able coloration in fishes'. *Amer. Natural.*, **69**, 245-66.

[87] SUMNER, F. B., 1935: 'Studies of protective color changes III. Experiments with fishes both as predators and prey'. *Proc. Nat. Acad. Sci.*, Washington, **21**, 345-53.

[88] SZYMANSKI, J. S., 1913: 'Ein Versuch, die für das Liebesspiel charak-teristischen Körperstellungen und Bewegungen bei der Wein-bergschnecke künstlich hervorzurufen'. *Pflüger's Arch.*, **149**, 471-82.

[89] THORPE, W. H., 1951: 'The learning abilities of birds'. *The Ibis*, **93**, 1-52, 252-96.

[90] TINBERGEN, L., 1935: 'Bij het nest van de Torenvalk'. *De Lev. Nat.*, **40**, 9-17.

[91] TINBERGEN, L., 1939: 'Zur Fortpflanzungsethologie von Sepia offici-nalis L.'. *Arch. néerl. Zool.*, **3**, 323-64.

[92] TINBERGEN, N., 1931: 'Zur Paarungsbiologie der Flusseeschwalbe (*Sterna h. hirundo L.*)'. *Ardea*, **20**, 1-18.

[93] TINBERGEN, N., 1935: 'Field observations of East Greenland birds I. The behaviour of the Red-necked Phalarope (*Phalaropus lobatus L.*) in spring'. *Ardea*, **24**, 1-42.

[94] TINBERGEN, N., 1936: 'The function of sexual fighting in birds; and problem of the origin of territory'. *Bird Banding*, **7**, 1-8.

[95] TINBERGEN, N., 1937: 'Über das Verhalten kämpfender Kohl-meisen (*Parus m. major L.*)'. *Ardea*, **26**, 222-3.

[96] TINBERGEN, N., 1939: 'Field observations of East Greenland birds II. The behavior of the Snow Bunting (*Plectrophenax nivalis sub-nivalis A. E. Brehm*) in spring'. *Trans. Linn. Soc. N.Y.*, **5**, 1-94.

[97] TINBERGEN, N., 1940: 'Die Übersprungbewegung'. *Zs. f. Tier-psychol.*, **4**, 1-40.

[98] TINBERGEN, N., 1942: 'An objectivistic study of the innate behaviour of animals'. *Biblioth. biotheor.*, **1**, 39-98.

[99] TINBERGEN, N., 1948: 'Social releasers and the experimental method required for their study'. *Wilson Bull.*, **60**, 6-52.

[100] TINBERGEN, N., 1950: 'Einige Beobachtungen über das Brut-verhalten der Silbermöwe (*Larus argentatus*)'. In: *Ornithologie als Biologische Wissenschaft, Festschrift E. Stresemann*, 162-7.

[101] TINBERGEN, N., 1951: *The Study of Instinct*. Oxford.

[102] TINBERGEN, N., 1951: 'On the significance of territory in the Herring Gull'. *The Ibis*, **94**, 158–9.

[103] TINBERGEN, N., 1951: 'A note on the origin and evolution of threat display'. *The Ibis*, **94**, 160–2.

[104] TINBERGEN, N., 1952: 'Derived activities; their causation, function and origin'. *Quart. Rev. Biol.*, **27**, 1–32.

[105] TINBERGEN, N., 1953: *The Herring Gull's World*. London.

[106] TINBERGEN, N., and VAN IERSEL, J. J. A.: Unpublished work.

[107] TINBERGEN, N., and KUENEN, D. J., 1939: 'Über die auslösenden und die richtunggebenden Reizsituationen der Sperrbewegung von jungen Drosseln'. *Zs. f. Tierpsychol.*, **3**, 37–60.

[108] TINBERGEN, N., MEEUSE, B. J. D., BOEREMA, L. K., and VAROSSIEAU, W. W., 1942: 'Die Balz des Samtfalters, *Eumenis* (= *Satyrus*) *semele* (*L.*)'. *Zs. f. Tierpsychol.*, **5**, 182–226.

[109] TINBERGEN, N., and MOYNIHAN, M., 1952: 'Head-flagging in the Black-headed Gull; its function and origin'. *Brit. Birds*, **45**, 19–22.

[110] TINBERGEN, N., and PELKWIJK, J. J. TER, 1938: 'De Kleine Watersalamander'. *De Lev. Nat.*, **43**, 232–7.

[111] TINBERGEN, N., and PERDECK, A. C., 1950: 'On the stimulus situation releasing the begging response in the newly hatched Herring Gull chick (*Larus a. argentatus Pontopp.*)'. *Behaviour*, **3**, 1–38.

[112] VERWEY, J., 1930: 'Einiges über die Biologie Ostindischer Mangrove krabben'. *Treubia*, **12**, 169–261.

[113] VERWEY, J., 1930: 'Die Paarungsbiologie des Fischreihers'. *Zool. Jahrb. Allg. Zool. Physiol.*, **48**, 1–120.

[114] WELTY, J. C., 1934: 'Experiments in group behaviour of fishes'. *Physiol. Zool.*, **7**, 85–128.

[115] WHEELER, M. W., 1928: *The Social Insects*. London.

[116] WILSON, D., 1937: 'The habits of the Angler Fish, *Lophius piscatorius L.*, in the Plymouth aquarium'. *J. Mar. Biol. Ass. U.K.*, **21**, 477–96.

[117] WINDECKER, W., 1939: '*Euchelia* (= *Hypocrita*) *jacobaeae L.* und das Schutztrachtenproblem'. *Zs. Morphol. Oekol. Tiere*, **35**, 84–138.

[118] WREDE, W., 1932: 'Versuche über den Artduft der Elritzen'. *Zs. f. vergl. Physiol.*, **17**, 510–19.

[119] WUNDER, W., 1930: 'Experimentelle Untersuchungen am dreistachlichen Stichling (*Gasterosteus aculeatus L.*) während der Laichzeit'. *Zs. Morphol. Oekol. Tiere*, **14**, 360–400.

INDEX